Smart Card Applications

Smart Card Applications

Design Models for using and programming smart cards

Wolfgang Rankl
Giesecke & Devrient GmbH, Germany

Translated by
Kenneth Cox
Kenneth Cox Technical Translations, Wassenaar, The Netherlands

BICENTENNIAL

1807

⊛WILEY

2007

BICENTENNIAL

Contents

Foreword

There was a tremendous breakthrough mood in the smart card world in the mid-1990s. The technology was seen to have reached a sufficient level of maturity and achieved sufficient functionality to enable a wide variety of security applications to be effectively implemented. The largest application areas were electronic purse systems – with an astonishing wealth of variants – and mobile communication systems, which were spreading over the entire world.

Unfortunately, system operators were confronted with many problems after large numbers of smart cards hosting these new, technically interesting smart card applications found their way into the hands of end users. There were instances in which no terminals were available for use by customers, and in some cases, system developers overlooked the fact that customers have their own interests and needs and cannot be manipulated to behave in a way that makes no sense to them.

Smart card technology has continued to develop unobtrusively in the meantime, and a paradigm shift has occurred in parallel with this development. Technology has vanished into the background as a driver for smart card applications, and its role as a guide to the future has been taken over by the applications. User needs now occupy the focus of attention. This is quite a normal cycle in the course of technology development, as has been seen repeatedly in this form and in similar forms.

This new aspect of the situation inspired me to write this book, as the current trend is to use smart cards on account of their positive characteristics instead of simply because they exist.

My objective with this book is not to elaborate on the theoretical aspects of abstract design models, but instead to concentrate on useful, proven solutions that can be implemented directly using available smart card operating systems. More than 15 years of professional experience with smart cards and their applications, as well as hundreds of letters from readers I have received as one of the co-authors of the *Smart Card Handbook*, have contributed to the creation of this book.

The design models described here are illustrated by a large number of interesting examples in order to maintain contact with real life. I have also intentionally included examples of failed and otherwise unsuccessful projects, because such examples often serve as excellent guides on how to do things better.

The central aim of this book is to describe reusable model solutions and modules that can be used to handle commonly occurring tasks and can be presented independently of actual program code. This is fully in accordance with the established method of breaking down a problem into smaller, subordinate problems that are easier to solve, developing individual solutions to these subordinate problems, and then combining the individual solutions to create an overall solution to the original problem.

This book is neither a reformatted version of the *Smart Card Handbook* nor an abridged version of that book, but instead a book that stands on it own and focuses on the subject of smart card applications. The first two chapters provide a brief introduction to the world of smart cards, but they address the underlying technology only to the extent necessary for a proper understanding of the following chapters. If you are interested in delving further in the technical details at any point, I take the opportunity here to refer you to the *Smart Card Handbook*.

I would like to express my thanks to the following people: Dieter Weiß for frequent and long discussions on the interpretation of ISO standards, Ralf Holly and Martin Rösner for many helpful tips on programming Java cards, Christoph Schiller for convincing me to use LaTeX, Sylvia Bernecker for the wonderful griffin, which looks just as I always imagined it but could never manage to realize on paper, Kenneth Cox for the translation, and of course Alexandra Rankl for her patience, without which I could never have written this book.

Munich, Spring 2006

Wolfgang Rankl
Rankl@gmx.net
www.WRankl.de

Symbols and Notation

– The least significant bit is designated as bit 1 in conformance with ISO nomenclature.

– In concatenated data elements, the higher-order byte is located at the start of the string and the lower-order byte is located at the end – the data format is thus big endian.

– The term 'byte' corresponds to its meaning in common usage and means a series of eight bits.

– The lengths of data elements and objects and all countable quantities are stated in decimal notation.

– When used in connection with data or memory sizes, the prefixes 'kilo', 'mega' and 'giga' have the values 1024 (2^{10}), 1 048 576 (2^{20}), and 1 073 741 824 (2^{30}) respectively. Similarly, the symbols 'KB', 'MB', and 'GB' designate 1024, 1 048 576, and 1 073 741 824 bytes.

– Binary values are used in a context-dependent manner and are not always explicitly identified as such.

– Smart card commands are set in upper-case letters (e.g. SELECT FILE).

– As a rule, only the positive results are shown in sequence charts.

Representation of Characters and Numbers

0, 1	Binary value (used according to context)
8	Decimal value
'00'	Hexadecimal value
"ABC"	ASCII value
bn	Bit number n (e. g. b8)
Bn	Byte number n (e. g. B1)
Dn	Digit number n (e. g. D3)

Logical Functions and Program Code

=	Assignment operator (also used as a comparison operator depending on the context)
$=, \neq, <, >, \leq, \geq$	Comparison operators

$+, -, \cdot, /$	Arithmetic operators
$\|$	Concatenation operator (e.g. for two data elements)

References

See '...'	This is a reference to another location in the book.
(N Y) *or* N (Y)	This is a reference to a document or Internet site listed in the bibliography. For documents with identified authors, 'N' is the last name of the first author listed in the bibliography and 'Y' is the year of publication. References to Internet sites and documents without identified authors are generally shown as unique abbreviations or organization names without a year.

Functions

$e = C(m)$	Calculate the error detection code e of the message m.
$t = T(d)$	Structure data d using TLV coding. The result is the TLV-coded data t.
$p = P(d, v, l)$	Pad data d to a integer block length l using the value or method v. The result is the padded data p.
$c = E(p, k)$	Encrypt plain text p using a symmetric cryptographic algorithm and the key k. The result is the cipher text c.
$p = D(c, k)$	Decrypt cipher text c using a symmetric cryptographic algorithm and the key k. The result is the plain text p.
$a = M(m, k)$	Calculate the message authentication code (MAC) of the message m using the secret key k.
$s = S(m, sk)$	Sign the message m using the secret key sk.
$r = V(m, s, pk)$	Verifying the signature s of the message m using the public key pk. The result is 'true' or 'false'.
$h = H(m)$	Calculate the hash value h of the message m.
$C = (A, pk_A, S(A \| pk_A, sk_{CA}))$	Generate the certificate C of the public key pk_A of user A. The certificate is signed using the secret key sk_{CA} of the certification authority CA.
$r = V(A \| pk_A, C, pk_{CA})$	Verify the certificate C of the public key pk_A of user A using the public key pk_{CA} of the certification authority CA. The result is 'true' or 'false'.

Abbreviations

3DES	Triple DES (data encryption standard)
3GPP	3rd Generation Partnership Project
ADK	additional decryption key
ADN	abbreviated dialling number
AES	Advanced Encryption Standard
AID	application identifier
API	application programming interface
ARM	Advanced RISC Machine
ARR	access rule reference
ASCII	American Standard Code for Information Interchange
ASN.1	Abstract Syntax Notation One
AT	attention
ATR	answer to reset
AUX1, AUX2	Auxiliary 1, Auxiliary 2
BAFA	Bundesamt für Wirtschaft und Ausfuhrkontrolle (German Federal Office of Economics and Export Control)
BCD	binary coded digit
BNA	Bundesnetzagentur (German Federal Network Agency)
BSI	Bundesamt für Sicherheit in der Informationstechnik (German Federal Office for Information Security)
CCS	cryptographic checksum
CDMA	code division multiple access
CEN	Comité Européen de Normalisation (European Committee for Standardization)
CHV	card holder verification information
CICC	contactless integrated chip card
CLA	class
CLK	clock
CPU	central processing unit
CRC	cyclic redundancy code
DES	Data Encryption Standard
DF	dedicated file
DO	data object
DPA	differential power analysis

DSA	digital signature algorithm
DSS	Digital Signature Standard
EC	elliptic curve crypto algorithm
ECC	elliptic curve cryptosystem
ECC	error correction code
ECDSA	elliptic curve digital signature algorithm (DSA)
EDC	error detection code
EEPROM	electrical erasable program read-only memory
EF	elementary file
EMV	Europay MasterCard Visa
ETSI	European Telecommunications Standards Institute
etu	elementary time unit
GND	ground (electrical)
GNU	GNU is not Unix
GPL	GNU General Public License
GUI	graphical user interface
HMAC	keyed-hash message authentication code (MAC)
HTML	hypertext markup language
I/O	input/output
IBE	identity-based encryption
ICAO	International Civil Aviation Organization
ICC	integrated chip card
ID	identifier
IEC	International Electrotechnical Commission
IFD	interface device
IMSI	international mobile subscriber identity
INS	instruction
IPR	intellectual property rights
ISO	International Organization for Standardization
ITU	International Telecommunications Union
JC	Java Card
JCP	Java Community Process
JCRE	Java Card runtime environment
JIT	just in time
JSR	Java specification request
L_c	length command
L_e	length expected
MAC	message authentication code
MD5	Message Digest Algorithm 5
MF	master file
MIPS	microprocessor without interlocked pipeline stages
NOP	no operation
NPU	numeric processing unit

NVM	nonvolatile memory
OCF	open card framework
OCR	optical character recognition
P1, P2, P3	Parameter 1, Parameter 2, Parameter 3
PC/SC	Personal Computer/Smart Card
PCD	proximity coupling device
PGP	Pretty Good Privacy
PIN	personal identification number
PIX	proprietary application identifier extension
PKI	public key infrastructure
PPS	Protocol Parameter Selection
PUK	personal unblocking number
RACE	Research and Development in Advanced Communications Technologies in Europe
RAM	random access memory
Reg TP	Regulierungsbehörde für Telekommunikation und Post (German regulatory agencies for telecommunication and postal services)
RF	radio frequency
RFC	Request For Comment
RFID	radio frequency identifier
RFU	reserved for future use
RID	registered application provider identifier
RIPEMD	RACE Integrity Primitives Evaluation Message Digest
RISC	reduced instruction set computer
RMI	remote method invocation
RND	random number
ROM	read-only memory
RSA	Rivest, Shamir and Adleman cryptographic algorithm
RST	reset
SAT	SIM Application Toolkit
SATSA	Security and Trust Services API
SECCOS	Secure Chip Card Operating System
SFI	short file identifier
SIM	subscriber identity module
SMS	short message service
SPA	simple power analysis
SPU	standard or proprietary use
SSC	send sequence counter
TDES	Triple DES (data encryption standard)
TETRA	Trans-European Trunked Radio
TLV	tag length value
TSCS	The Smart Card Simulator
UART	universal asynchronous receiver transmitter
UCS	universal character set

UICC	universal integrated chip card
UML	unified modelling language
UMTS	Universal Mobile Telecommunication System
USB	Universal Serial Bus
USIM	universal subscriber identity module
Vcc	supply voltage
VM	virtual machine
XML	extensible markup language
XOR	logical exclusive OR operation

Chapter 1

Overview of Smart Cards

In contrast to information technology practices in the PC realm, the development and functionality of smart cards are strongly driven by international standards. The reason for this is that interoperability and interchangeability are very important factors for smart cards. From the very beginning, this has fostered specification of their characteristics in standards. Another significant factor is that none of the suppliers of smart card hardware or software has ever held a monopoly position.

1.1 Card Classification

If you were to classify smart cards in the same manner as living beings in biology, you would obtain a tree chart similar to what is shown in Figure 1.1. The top level includes all types of cards, which can have various formats.

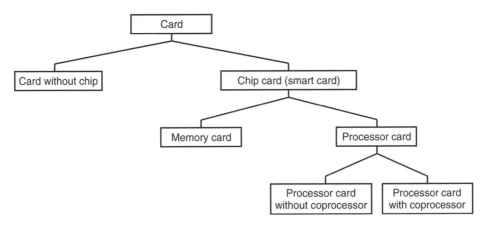

Figure 1.1 Classification of cards with and without chips

Smart Card Applications: Design Models for using and programming smart cards   W. Rankl
© 2007 John Wiley & Sons, Ltd

Cards can be divided into cards without chips and cards with chips. Logically enough, the latter type are called *chip cards*, which are also commonly known as *smart cards*. The chip, which is the essential distinguishing element, can be either a memory chip, in which case the card is called a *memory card*, or a *microcontroller chip*, in which case the card is called a *processor card*. Processor cards can be further subdivided into processor cards with or without coprocessors for executing asymmetric cryptographic algorithms such as RSA (Rivest, Shamir and Adleman) or ECC (elliptic curve cryptosystems).

This classification provides an adequate overview of the most widely used types of cards. However, it can also be extended to include devices that use smart card technology. The best-known examples of such devices are 'super smart cards' and tokens. A super smart card has a direct user interface to the smart card microcontroller, in the form of additional card elements such as a display and buttons. A token has a different form that is better suited to its intended use than the usual card format. Typical examples include tokens in the form of USB plugs that can be connected directly to a PC. However, the underlying technology is still the same as that of smart cards, with only the appearance being different.

1.2 Card Formats

The most common types of cards in current use have one feature in common, which is a thickness of 0.76 mm. As illustrated in Figure 1.2, all other dimensions can differ. These formats are not arbitrary. Instead, they are specified by international standards or by specifications stipulated by major card issuers. This is also important, since at least in case of contact cards they must be able to fit into corresponding terminals or readers.

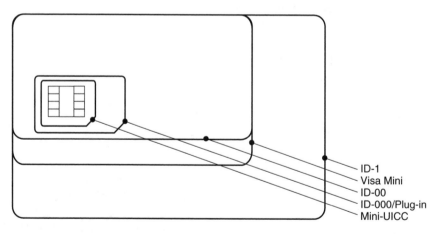

ID-1
Visa Mini
ID-00
ID-000/Plug-in
Mini-UICC

Figure 1.2   Relative sizes of commonly used card formats

Typical smart card formats are summarised in Table 1.1. The most commonly used card format, which is also undoubtedly the best known format, is ID-1. The reason it is so widely used is that practically all credit cards and other forms of payment cards are made

Table 1.1 Summary of typical card formats. All stated dimensions are exclusive of tolerances. All formats have the same thickness: 0.76 mm

Card Format	Width (mm)	Height (mm)	Use
ID-1	85.6	54	Well-known standard format
ID-00	66	33	Standardized for telecommunications, but not used
Visa Mini	65.6	40	Payment systems
Plug-in, ID-000	25	15	Telecommunications
Mini-UICC	15	12	Telecommunications

in this format. The plug-in format for smart cards used in mobile telecommunications applications is also very common. Another name for this format is ID-000. This has become the standard format for cards used in mobile telephones.

The recently defined mini-UICC format is also available for the mobile telecommunications sector. It was developed in response to the ongoing miniaturization trend that prevails in this sector. The Visa Mini format is a smaller version of the ID-1 format. It is intended to meet customer demand for cards with the smallest possible dimensions.

Cards with shapes other than the usual rectangular card body are also being made now. For example, there are cards with one corner rounded at a large radius and cards shaped in the outline of an animal. The constraints with respect to the shape of contact cards are that they must fit into the slot of an ID-1 terminal, be readily removed from the terminal after use, and make reliable electrical contact with the terminal. Incidentally, most cards with special shapes are made by stamping them from cards in ID-1 format to achieve the desired shape.

1.3   Card Elements

The card body is usually more than just a carrier for the chip module. It also includes information for the user and card accepters and of course security elements for protection against forgery. Furthermore, the card body is an excellent advertising medium. The card issuers must coordinate all these functions, some of which are mutually contradictory, with their own specific wishes. The ultimate result is the issued card.

1.3.1   Printing and labelling

A rather wide variety of processes are available for printing and labelling cards. Text elements that are common to all cards of a series are normally applied using offset printing or silkscreen printing, but sheet printing and individual card printing processes are also used.

Lasering is widely used for printing individual cards. This consists of using a laser beam to darken the surface of the plastic card body. This process produces irreversible card labelling, but it requires a certain amount of investment in technology. A more economical alternative is thermal transfer printing, which can also be used for colour printing. One of the drawbacks of this method is that the colour layers are located close to the surface of the card, so they can be removed almost completely. Digital printing processes for high-quality printing of individual cards are a relatively new development.

1.3.2   Embossing

The main advantage of embossing, which is commonly used with credit cards, is that the labelling can be transferred to paper using a simple stamping machine. The embossed section of the card can be restored to its original state by heating the card to a relatively high temperature. For this reason, the check digits at the end of the embossing usually extend into the hologram area. As the hologram will be visibly damaged if the card is heated, this makes it relatively easy to detect manipulation of the embossing.

1.3.3   Hologram

Technically sophisticated equipment is necessary to produce the white-light reflection holograms used on cards. As forgers usually do not have access to such equipment, holograms are commonly used on smart cards as security features. Some other reasons for using holograms are that they are inexpensive in large quantities, they can be checked directly by users, and the hologram cannot be removed from the smart card without destroying it. Unfortunately, there is no link between the hologram and the microcontroller, which reduces its advantages from the perspective of the chip.

1.3.4   Signature panel

The signature panel is located on the rear of the card. It must be erasure-proof so that the signature on the panel cannot be removed without it being noticed. A coloured pattern is often printed on the signature strip, so any attempt to manipulate the signature will cause visible damage to the pattern.

1.3.5   Tactile elements

Tactile elements can be applied to the card to enable visually impaired and blind people to recognize the orientation of the card. The best known example is a semicircular recess in one of the long edges of the card. The hole punched in some payment cards is also suitable for use as an orientation aid, although its original purpose was to allow the card to be hung from a strap or cord.

1.3.6   Magnetic stripe

With many types of cards, the only reason to retain the magnetic stripe (with its data storage capacity of a few hundred bytes) is compatibility with a widely distributed terminal infrastructure. However, it will still take a long time before magnetic-stripe cards are fully replaced by smart cards, since they are significantly cheaper.

1.3.7   Chip module

The chip module is a protective housing for the microcontroller chip, which is fitted to the rear of the module. The module can have six or eight visible contacts on its external surface, although modern smart cards need only five contacts. The other contacts are reserved for future applications. The microcontroller is glued to the rear of the contact substrate and electrically connected to the contact surfaces on the front side by thin bonding wires. Figure 1.3 shows the signal assignment of the contacts of a chip module.

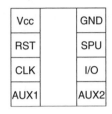
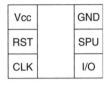

Figure 1.3   Contact assignments of a smart card module. Abbreviations: Vcc = Supply voltage, RST = Reset, CLK = Clock, AUX1 = Auxiliary 1, GND = Ground, SPU = Standard or Proprietary Use, I/O = Input/Output, AUX2 = Auxiliary 2

1.3.8   Antenna

Smart cards that communicate without using contacts must have an integrated antenna in the card body. The antenna is a sort of coil consisting of several turns along the outer edge of the entire card. Various methods can be used to produce the antenna. Methods that are used in practice include a coil of thin copper wire embedded in the card body, etched copper tracks, and printed coils.

1.4   Smart Card Microcontrollers

The characteristics of a smart card are largely determined by its microcontroller. Single-chip microcontrollers are normally used. A single-chip microcontroller consists of a small silicon chip equipped with all the functions necessary for its intended use. Smart card microcontrollers are not standard microcontrollers such as those used in coffee machines and toasters, but are instead chips specially adapted for use in smart cards. The adaptations encompass electrical and physical parameters such as the maximum current consumption, the range of allowed clock frequencies, and the allowable temperature range.

Besides all these functional parameters, there is another essential item: security functions. Smart card microcontrollers are specially hardened against attacks. This includes detecting undervoltage and overvoltage conditions and detecting clock frequencies outside the specified range. These microcontrollers also incorporate light and temperature sensors to enable them to recognize attacks via these routes and respond accordingly.

However, these are only relatively simple protective mechanisms. There are also relatively complex methods, which are quite widely used, such as scrambling all the memories and the busses between the processor and the memories. It is even possible to periodically swap the scrambling key during an individual session. The microcontroller hardware can even defend against hard attacks such as measuring its current consumption in order to perform a statistical analysis to discover which data was processed by the processor.

Besides technologically advanced smart card microcontrollers, there are also memory chips which are essentially intended to be used as simple data storage devices with fixed logic circuitry designed by the semiconductor manufacturer. Figure 1.4 shows the basic functional groups present on the chip. The ROM (read-only memory) contains data about the chip type. The EEPROM (electrically erasable programmable read-only memory) provides the storage area for a unique chip identification number and data stored in read/write memory. A terminal can store several hundred bytes to a few thousand bytes of data here.

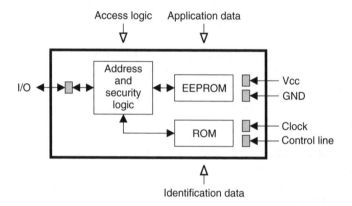

Figure 1.4   Block diagram of a memory chip for a smart card with a contact interface

The security logic, which varies according to the chip type, monitors access to the data. For instance, successful verification of a PIN (personal identification number) in the memory chip may be necessary before write access is possible.

Telephone cards, which are chip cards that can be used with public pay phones, have a similar operating principle. The security logic of a telephone card incorporates an authentication algorithm so that the telephone can determine whether it is dealing with a genuine chip card. If the card is genuine, a counter in the EEPROM is decremented according to the duration of the call. This counter can only count down, and it stops when it reaches zero. When this happens, the card has been used up.

Microcontrollers for smart cards have significantly more functionality than simple memory chips, as can be seen from Figure 1.5 on the facing page. The CPU (central processing unit) is a freely programmable control unit that executes the machine instructions of the operating system, which is located in the ROM. The CPU is assisted by a numerical coprocessor (NPU – numeric processing unit) for numerical calculations, particularly those dealing with cryptography. These special processors combine extremely high performance with low power consumption. Operating system extensions and the actual

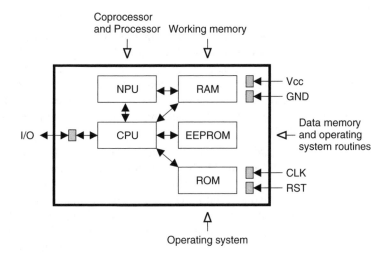

Figure 1.5 Block diagram of a microcontroller for a smart card with a contact interface

applications and associated data are stored in the EEPROM. Just as in a PC, the RAM (random-access memory) serves as working memory to hold data during operation.

These functional groups must all be integrated in a single chip that is limited to a maximum size of 25 mm² for reasons of strength and robustness. As a consequence, the amount of available memory is many orders of magnitude less than what is commonly found in a modern PC. The ROM capacity of smart card microcontrollers typically ranges from 16 to 400 KB, the EEPROM capacity ranges from 1 to 500 KB, and the RAM size ranges from 256 bytes to 16 KB. These wide ranges are due to the wide variety of application areas. The simplest processor cards do not even have an operating system, but instead contain only the application software. At the other extreme, smart cards currently at the top of the technology ladder fully exploit all the available memory.

These memory sizes are quite normal in the embedded applications area, but they are mini-memories compared with the memories of modern PCs. Nevertheless, the semi-conductor technology of smart card microcontrollers is comparable to the technology used to manufacture modern high-performance processors, since integrating the vari-ous memory technologies and the necessary hardening against attacks is rather difficult. The microcontrollers are fabricated using semiconductor processes with 90-nm technol-ogy, which is only one development step away from the current state-of-the-art 65-nm technology.

Additional interfaces are integrated into smart card microcontrollers to expand their range of potential uses. For instance, the commonly used half-duplex bit-serial port can be augmented by a USB interface or a wireless communication interface. Semiconduc-tor manufacturers usually base such developments on existing smart card microcon-trollers, which are upgraded to support the additional interfaces. The result is thus a single-chip microcontroller that can communicate with the outside world via additional interfaces.

1.4.1 Processor

If you analyse the sales volumes of currently used smart card microcontrollers, you will find that most of them still have an 8-bit CPU. This is usually a simple 8051 CPU, which has proved itself over the last two decades, along with a few extensions. The processing power of such a CPU is sufficient for all operating systems that do not include an interpreter. However, if the operating system must provide a Java interpreter, there is a distinct preference for microcontrollers with 16-bit processors. Some of these processors are also based on a modified 8051 architecture.[1]

There are also a few smart card microcontrollers that are based on well-known 32-bit processor families such as ARM 7 or MIPS. The limiting factor for using such high-performance processors is the chip area. There is a more or less direct relationship between chip area and price, and a 32-bit processor occupies a significantly larger area than an 8-bit processor. It is often more economical to invest in optimizing the speed of the software than to use a processor that needs more chip area. This is ultimately a consequence of the fact that smart cards have to be low-cost, mass-production items.

1.4.2 Memory

In addition to a processor, every microcontroller needs several types of memory with differing characteristics. The main type of nonvolatile memory used in smart card microcontrollers is ROM. If the data located in memory must be modified in operation, electrically erasable memory (EEPROM) is used.

Besides microcontrollers with ROM and EEPROM, a steadily increasing number of chips with flash memory are being used. Flash memory is a sort of EEPROM with reduced cell dimensions, but unlike EEPROM it cannot be erased or written byte-wise. Flash memory can take over the functions of ROM and EEPROM.

EEPROM and flash memory are similar in that they cannot be erased and written an unlimited number of times and these accesses cannot occur at the full speed of the processor. Currently, the erase and write times are typically 3.5 ms each, and the guaranteed number of such accesses is 500 000. This has a major impact on the design of the operating system and application software.

Static RAM is used as volatile memory for storing data during operation.

1.4.3 Supplementary hardware

Besides a processor and its associated memory, smart card microcontrollers incorporate various types of supplementary hardware. Figure 1.6 shows a large range of possibilities.

The clock signal required by the smart card is usually provided by the terminal. However, as the relevant standards restrict the frequency of this clock signal to a range of 1–5 MHz, more and more microcontrollers include internal clock multiplier or clock generator circuitry.

---

[1] If you are willing to generate the time-critical parts of the operating system in assembly language instead of C and invest a fairly significant amount of time in optimizing its real-time behaviour, it is certainly possible to develop an interpreter that will run at an acceptable speed on an 8-bit CPU

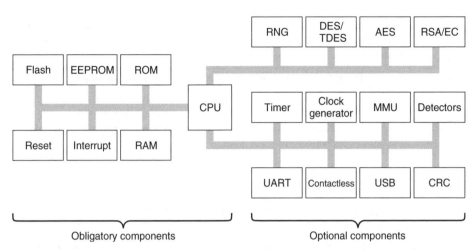

Figure 1.6 Block diagram of a smart card microcontroller with a selection of currently common components linked to the CPU via a shared address, data and control bus. The ROM and EEPROM memories may be omitted in some types of microcontrollers if flash memory is used

A UART (universal asynchronous receiver transmitter) is included for bit-serial communication with the terminal, and in the case of smart cards with USB or contactless interfaces, the corresponding communication components are also present in the hardware.

Most of the supplementary hardware is related to cryptography, since considerable processing power is sometimes necessary for this purpose. Random numbers are almost always generated using a hardware random number generator, although the results are further processed in software before being used. Symmetrical cryptographic algorithms such as DES (Data Encryption Standard), Triple DES (TDES) and AES (Advanced Encryption Standard) are also usually present in hardware, and they generally require only a few clock cycles for full encryption or decryption.

Hardware for computing asymmetric cryptographic algorithms is not generally included in all microcontrollers, as it would increase the price. If it is present, it supports the usual algorithms such as RSA (Rivest, Shamir and Adleman cryptographic algorithm), DSA (digital signature algorithm) and ECC (elliptic curve cryptosystems). The hardware implementation of such algorithms is always kept relatively modular to enable it to support various key lengths and versions, extending as far as key generation.

1.4.4 Electrical characteristics

In mobile telecommunication applications, low power consumption of all components of a mobile telephone is a visible feature even for end users, since it directly affects the speech and standby times of the telephone. The mobile telecommunication sector has thus developed into a driver for smart cards with the lowest possible operating voltages and current consumptions. This is in full contrast to all terminals connected directly to

Table 1.2   Voltage classes as specified by ISO/IEC 7816-3. The stated maximum clock rate is a typical value, which can optionally be changed to a wider range (4–20 MHz). The terminal must be informed of this via the ATR

Voltage Class	Voltage	Clock Frequency	Current Consumption
Class A	5 V (±10 %)	1–5 MHz	60 mA maximum at maximum clock frequency
Class B	3 V (±10 %)	1–5 MHz	50 mA maximum at maximum clock frequency
Class C	1.8 V (±10 %)	1–5 MHz	30 mA maximum at maximum clock frequency

the mains network, for which the current consumption of the smart card is practically an insignificant issue.

Voltage class A, with a 5-V supply voltage and (originally) a maximum allowable current consumption of 200 mA, has completely disappeared in mobile telephone applications. The current state of the art is still voltage class B, with a clearly visible trend in the direction of the 1.8-V C class, as summarized in Table 1.2. On the other hand, 5-V smart cards are still commonly used in payment systems.

Incidentally, modern smart cards can usually work with all three voltage classes. However, the processing power may decrease with decreasing supply voltage. This is due to the internal frequency multiplication of the chip, which depends on the amount of power available.

Chapter 2

Smart Card Operating Systems

The nature of a smart card depends more on the operating system running in the card than on the microcontroller implanted in the card. The operating system is what transforms a piece of plastic with an embedded processor, memory and a few peripheral functions into a full-fledged smart card with its presently known range of uses.

Current smart card operating systems are stored in the ROM of the microcontroller in unalterable form. They use a large portion of the available RAM and a small portion of the EEPROM. Nearly all commonly used smart card operating systems are based on the provisions of the ISO/IEC 7816 family of standards.

Smart card operating systems can be classified into native operating systems and interpreter-based operating systems. Native smart card operating systems and the applications that run under them execute in the machine language of the associated target processor. They are usually generated in the C programming language, and they do not have an interpreter or compiler to translate programs into the machine language of the target processor.

Most interpreter-based operating systems are also written in C, but the application programs that run under them do not have to be generated in the machine language of the target processor. Instead, they can be written in an interpreted programming language such as Java. Consequently, these operating systems incorporate an interpreter to translate programs into the machine language of the target processor. Some well-known examples of interpreter-based operating systems are Java Card, BasicCard, and Multos.

2.1 File Management

Managing files is the principal task of a smart card operating system. File management means not only providing read and write access to files and creating and deleting files, but also granting access privileges and monitoring compliance with access privileges. File management is especially important because most smart card applications are file-based.[1]

[1] See Section 3.3.2

Smart Card Applications: Design Models for using and programming smart cards   W. Rankl
© 2007 John Wiley & Sons, Ltd

File management in smart cards is almost entirely based on the provisions of the ISO/IEC 7816-4 standard. They specify a maximum possible functional scope, which in turn is implemented in actual smart card operating systems only to the extent necessary.

2.1.1   File types

Smart card file structures are always based on a tree structure with a root directory, as illustrated in Figure 2.1. The root directory of a smart card, which is analogous to the 'c:' volume of a PC, is called the *MF* (*master file*) and is present only once in the file tree of the smart card. It has the properties of a directory, which means it can only contain other directories and cannot store data directly.

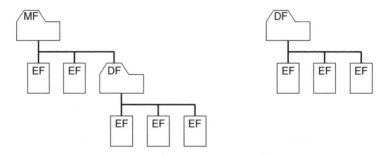

Figure 2.1   The two possible forms of file-based applications in smart cards. A simple smart card file system is shown on the left. It contains an MF with application-independent EFs located directly below the MF, along with a DF with application data contained in EFs. A DF without a visible MF is shown on the right. It also contains application data in the form of EFs located below the DF. This sort of DF is also called an *ADF*

The directories of a smart card are called *DFs* (*dedicated files*), and in theory they can be nested indefinitely. Three or four levels are commonly used in actual applications, and smart card operating systems rarely support more than eight levels. The ADF (application dedicated file) is a special type of DF. It is a DF for a specific application and can be located in the file tree of the smart card without there being any direct relationship to the root directory. Typically, it holds all the files of a particular application. ADFs are rarely encountered in actual practice.

The actual application data and operating system data are stored in EFs. EFs are always located in directories, and there are two possible types: working EFs and internal EFs. Working EFs are used to store application data that is accessible to the outside world via smart card commands. By contrast, internal EFs are used by the smart card operating system to store data for internal purposes. For example, they can be used to store keys or a seed (initial value) for a random number generator.

2.1.2   File names

As smart cards are always used under the control of a terminal, it is not necessary to make the file names compatible with human needs. Standard file names thus consist of a 2-byte data element called the *FID* (*file identifier*). The FID of the MF, which is '3F00', is reserved for this purpose. All other FIDs can be freely chosen. Table 2.1 lists the file names of commonly used types of smart card files and summarises their key characteristics.

Each directory file (DF) has a supplementary name in addition to its FID, and it can be addressed in the file tree using this supplementary name. This supplementary name is called the *DF name*, and it usually includes an AID (application identifier). The AID consists of an RID (registered application provider identifier) and a PIX (proprietary application identifier extension). RIDs can be registered officially to ensure that they are unique throughout the world. In this case, the PIX can be used as necessary to further identify a specific DF. This makes it possible to define a unique name for a specific smart card application, which can then be used to recognize and select it in every smart card.

The EFs provided to hold data are also assigned FIDs, similar to all smart card files. In addition, each EF has an SFI (short file identifier), which can be provided as a parameter of a read or write command to select the EF directly.

Table 2.1   Possible file names as specified by ISO/IEC 7816-4. The restrictions on the range of values for the FID described in Section 5.3.2 must be observed

Data Type	File Name	Size	Value Range
MF (master file)	FID (file identifier)	2 bytes	'3F00'
DF (dedicated file)	FID (file identifier)	2 bytes	0 ... 'FFFF'
	DF name (usually includes an AID)	1–16 bytes	0 ... 'F ... F'
	AID (RID ‖ PIX)	5–16 bytes	According to AID definition
EF (elementary file)	FID (file identifier)	2 bytes	0 ... 'FFFF'
	SFI (short file identifier)	5 bits	1 ... '30'

2.1.3   File structures

Smart card data files (EFs) have internal structures. This means that the data stored in the files can be arranged in various ways. Five different structures are available, as illustrate in Figure 2.2.

In the transparent structure, the data items are arranged as a series of bytes (byte string). The commands READ BINARY and UPDATE BINARY can be used to read data from or write data to this file structure using parameters that specify an integral number of bytes and an offset from the start of the file. This EF structure is a general-purpose structure that can be put to a wide variety of uses.

The maximum file size is not specified, but the maximum address range of READ BINARY and UPDATE BINARY limits it to 33 023 bytes (consisting of a maximum offset of 32 768 bytes and a maximum read or write length of 255 bytes).

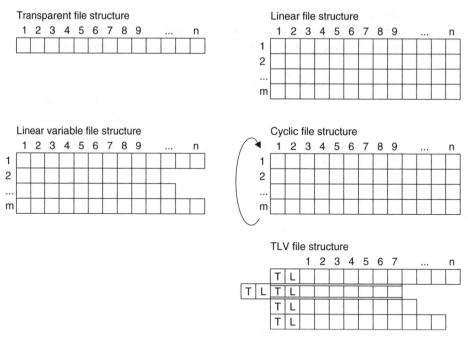

Figure 2.2   The five possible structures of data files (EFs) used in smart cards. Each cell in the diagrams represents a data byte

Besides the transparent file structure, there are three record-oriented file structures. EFs with a linear fixed file structure can be used to store equal-length records. The linear variable file structure allows the records to have different lengths. If records with different lengths must be stored in a smart card, the amount of memory space required will be less if a linear variable EF is used than if a linear fixed EF is used. These two file structures are typically used to store personal data such as addresses or telephone numbers.

The cyclic file structure extends the linear file structure to include a pointer that indicates which record was most recently written. This structure is thus ideal for a variety of log file applications.[1]

The records of all record-oriented files can be read and written using the READ RECORD and UPDATE RECORD commands. Normally, it is only possible to read or write complete records although relatively recent operating systems also support access to partial records.

The fifth type of file structure enables data objects to be stored in a TLV structure. In such a structure, each data object is identified by tag (T) and length (L) elements, which are followed by the actual data or value (V). This file structure can also be used to store nested data objects. Data objects can be read and stored using the GET DATA and PUT DATA commands.

Table 2.2 lists commonly used types of smart card files and summarises their key characteristics.

---

[1] See Section 5.4

Table 2.2   Reasonable minimum and maximum file sizes. The restrictions do not result directly
from any standards, but instead result indirectly from the limitations of the access
commands. The standards are vague on this subject, and sometimes even mutually
contradictory. Consequently, our recommendation here is to always maintain a safety
margin relative to the limits and in any case to make a preliminary test with the smart
card operating system you intend to use

File Structure	Typical File Size	
Transparent	Total size	1–33 023 bytes
Linear	Record length	1–255 bytes
	Number of records	1–254
Linear variable	Record length	1–255 bytes
	Number of records	1–254
Cyclic	Record length	1–255 bytes
	Number of records	1–254
TLV	Data object size	Not specified (typically 65 535 bytes)
	Number of data objects	Not specified (typically 255)

2.1.4   File attributes

Files in smart cards can also have various attributes, depending on the specific operating
system. The best-known set of attributes is *shareable* and *not shareable*. These attributes
can be used to specify for each file whether it permits concurrent read or write access
via multiple logical channels. There are many other possible file attributes, but they are
not standardized.

2.1.5   File selection

The smart card SELECT command is used to explicitly select a file. A file must always
be selected before it can be accessed with the usual commands such as READ BINARY
or UPDATE BINARY.

One of the available identifiers (FID, DF Name or AID) must be used for selection,
depending on the file type (MF, DF or EF). These identifiers do not have to be unique
in the directory and file structure of a smart card. Consequently, the selection options
depend on the currently selected file. Figure 2.3 illustrates the selection methods that
can normally be used in the directory and file structure.

Selection using a path name enables fast selection across several DFs with a single
command. With this method, the path to the file to be selected is passed to the smart
card as a command parameter. This path can be referenced to the MF or to the currently
selected file. This is the simplest selection option, and above all, it is the option that
requires the least amount of transaction time. The MF can be selected in a similar manner.
It can be selected from anywhere in the entire file tree using a single command.

The four commonly used read and write commands (READ BINARY, UPDATE BI-
NARY, READ RECORD and UPDATE RECORD) also support file selection during

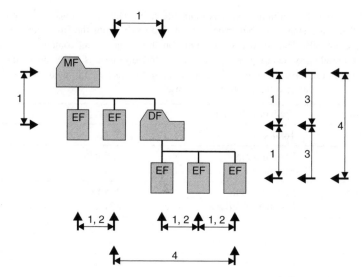

Figure 2.3   File selection options for smart cards. Option 1 is explicit selection using an FID (file identifier); option 2 is implicit file selection using an SFI (short file identifier); option 3 is selection using a DF name; option 4 is selection using an FID (file identifier) and a path parameter

command transaction (implicit selection). This eliminates the need to use SELECT to select the desired file before issuing the actual read or write command. This function is called *implicit file selection*, and it is quite useful for reducing file access times.

2.1.6   Access conditions

Access conditions associated with the files defined in a file system are an essential component of the file system. They specify which conditions must be satisfied to enable read or write access to the files. These conditions could be, for example, successful PIN verification or successful authentication of the terminal by the smart card.

Two different methods are commonly used in smart cards for technical implementation of access conditions: state-based access conditions and rule-based access conditions. The first method has been used for more than a decade in large systems, such as the SIMs used in GSM mobile telecommunication systems. Rule-based access conditions were first published as a standard[1] in the late 1990s. They are actually just a generalization and extension of the state-based method. As a result, all aspects of state-based access conditions can be reproduced using rule-based conditions.

2.1.6.1   State-based access conditions

In the case of state-based access conditions, each form of access (read or write) is only possible if a certain state has been attained, independent of other forms of access.

---

[1] See ISO/IEC 7816-4 (2005)

The EF$_{ADN}$ (abbreviated dialling number) file of a SIM can be used here as a typical example. This file can only be read using the READ RECORD command if PIN 1 has previously been correctly verified by the smart card.

Nearly all file-based smart card applications can be implemented with relative ease using state-based access conditions. However, a growing number of smart card operating systems support the rule-based method, which is more future-proof and significantly more flexible.

2.1.6.2 Rule-based access conditions

Rule-based access conditions in smart cards are based on assigning all files (DFs and EFs) references to a record-oriented file containing sets of access rules. This file is assigned the name EF$_{ARR}$ (access rule reference), and each reference is simply composed of the FID of the EF$_{ARR}$ and a record number that addresses the appropriate set of rules. The FID of EF$_{ARR}$ is freely selectable.

Each record in EF$_{ARR}$ contains a set of rules for the various forms of access, such as read and write. As directory files can also be assigned references to an EF$_{ARR}$, it is also possible to define rules for creating and deleting files.[1] This method can also be used in a similar manner to manage access to data objects.

With rule-based access conditions, it is even possible to specify that certain files can only be accessed using Secure Messaging.[2] The ISO/IEC 7816-9 standard forms the basis for the coding and the available functionality, but you should always consult the specifications of the smart card operating system being used, since the standard provides many options and there are large differences between individual operating systems. The operating principle of rule-based access is illustrated in Figure 2.4.

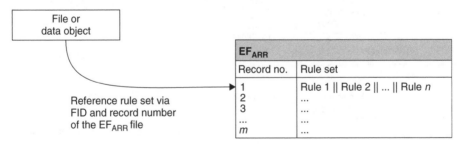

Figure 2.4   Operating principle of using an EF$_{ARR}$ to manage rule-based access conditions for files and data objects

All commonly encountered requirements for access to files and data objects in smart card applications can be implemented using rule-based access conditions. Although this method is not especially simple, it is very powerful. As a comment regarding security, we can note here that it is essential to ensure that write accesses to EF$_{ARR}$ can only be performed by authorized entities. Otherwise, the entire security of an application can be effectively bypassed.

---

[1] See Section 5.3
[2] See Section 2.3.4

A mistake in connection with EF_{ARR} that can nearly be regarded as classic must be mentioned here. If it is possible to freely delete and create files in the directory containing EF_{ARR}, the following simple but highly effective attack is possible. The attacker first uses DELETE to delete EF_{ARR} and then uses CREATE to create a new EF_{ARR} in which all read and write conditions for the files that reference this file are set to 'always'. After this, the attacker can use standard commands to read all EFs containing application data, and of course the attacker can also alter the contents of these files. Although this is essentially a primitive form of attack, it shows quite clearly that even a sophisticated method such as rule-based access requires suitably careful planning.

2.1.7   File life cycle

In the ideal case, it is possible to create, use and then delete files in a smart card file system whenever so desired. In addition, the amount of free memory available to the file system is ideally just as large after completion of this cycle as at the beginning. The life cycle of files, including all possible options, is illustrated in Figure 2.5.

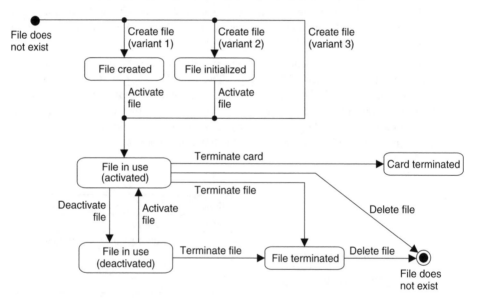

Figure 2.5   States and associated state transitions during the entire life cycle of a file, as specified by ISO/IEC 7816-9

All these options are actually available in large smart card operating systems. On the other hand, simple operating systems often have restrictions in this regard. For instance, simple operating systems often do not allow files to be deleted once they have been created or if they do allow files to be deleted, the amount of available free memory may be reduced by several bytes for each pass through the described life cycle.

Of course, these simple operating systems have the advantage that they can run on microcontrollers with significantly less processing power (and correspondingly lower prices) than what is required to run an operating system that supports the full range of options of the file life cycle. The simpler version is entirely adequate for many applications because, quite often, only the file contents are modified in actual practice, and never the actual files.

2.2   Commands

Aside from file management, commands are the most important functionality that a smart card operating system provides to the outside world. The number of commands supported by modern operating systems can easily amount to around 50. This is entirely adequate for implementing most applications, including complicated applications, without using supplementary user-defined commands. Table 2.3 provides a summary of standard smart card commands, and Table 2.4 provides a selected list of the most commonly encountered return codes sent by smart cards in response to commands received from a terminal. With regard to the exact coding of individual commands, you must always refer to the specifications of the smart card operating system being used.

Commands for file operations   The commands for file operations include SELECT, which is used to select a specific file, and READ BINARY and READ RECORD, which are used to read data from files having various structures. By contrast, UPDATE BINARY and UPDATE RECORD are the commands for writing data to files. The search commands SEARCH BINARY and SEARCH RECORD can be used to search for specific values in the EFs of the associated directory and file structure.

Commands for file management   The commands for file management are used for administrative purposes to manage the directory files (DFs) and data files (EFs) in the file tree of a smart card. This includes using CREATE FILE to create new files, APPEND RECORD to enlarge files, and DELETE FILE to delete existing files. The ACTIVATE FILE and DEACTIVATE FILE commands block and unblock files. The TERMINATE DF and TERMINATE EF commands permanently block files without deleting them from the file tree.

Commands for data objects   Application data can be stored in data objects and/or files. GET DATA and PUT DATA read data from data objects and write data to data objects.

Commands for security functions   The best-known security function command is VERIFY, which is used to verify PINs. GET CHALLENGE requests a random number for a subsequent EXTERNAL AUTHENTICATE command, which is used to authenticate the outside world with respect to the smart card. By contrast, INTERNAL AUTHENTICATE can be used to authenticate a smart card with respect to the rest of the world by using a challenge–response process. MUTUAL AUTHENTICATION can be used to authenticate the smart card and the outside world with respect to each other in a single operation.

The PERFORM SECURITY OPERATION (PSO) command can be used to invoke all the cryptographic functions of a smart card under the control of parameters passed

Table 2.3 Annotated list of the most important smart commands defined by ISO/IEC 7816-4, -8, -9, and Open Platform

Function Class	Command	Description
File	SELECT	Select a file operation
	READ BINARY READ RECORD	Read data from a transparent or record-oriented file
	UPDATE BINARY UPDATERECORD	Write data to a transparent file or record-oriented file
	SEARCH BINARY SEARCH RECORD	Search for a pattern in a transparent file or record-oriented file
File management	CREATE FILE	Create a file (DF or EF)
	APPEND RECORD	Create a new record in a record-oriented file
	ACTIVATE FILE	Reversibly unblock a file
	DEACTIVATE FILE	Reversibly block a file
	TERMINATE DF/EF	Permanently block a file (DF or EF)
	DELETE FILE	Delete a file (DF or EF)
Data objects	GET DATA	Read TLV-coded data objects
	PUT DATA	Write TLV-coded data objects
Security	VERIFY	Verify transferred data
	GET CHALLENGE	Request a random number (e.g. for a subsequent EXTERNAL AUTHENTICATE)
	INTERNAL AUTHENTICATE	Unilateral authentication of the smart card by the outside world
	EXTERNAL AUTHENTICATE	Unilateral authentication of the outside world by the smart card
	MUTUAL AUTHENTICATION	Mutual authentication of the smart card and the outside world
	PERFORM SECURITY OPERATION	Execute a cryptographic algorithm in the smart card
	MANAGE SECURITY ENVIRONMENT	Manage security command parameters
Program code management	LOAD	Load a code-based application
	INSTALL	Install a code-based application
	PUT KEY	Load a key for a code-based application
	SET STATUS	Write state information for the life cycle of the smart card or an application
	GET STATUS	Read state information about a security domain, load file or application
	DELETE	Delete an object
Data transmission	GET RESPONSE	Request data for the T=0 transmission protocol from the smart card

Table 2.4 Classification of return codes defined by ISO/IEC 7816-4. The 'Normal processing' and 'Warning processing' groups belong to the 'Process completed' category. The 'Execution error' and 'Checking error' groups belong to the 'Process aborted' category, for which only the return code is sent back to the terminal without any response body. The variable 'xx' can take on values in the range of '00' to 'FF'

Group	SW1 ‖ SW2	Meaning
Normal processing	'9000'	Process executed successfully
	'61xx'	Processing completed successfully. xx data bytes are available in response and can be retrieved using GET RESPONSE
Warning processing	'62xx'	Data in nonvolatile memory not modified. See SW2 for details
	'63xx'	Data in nonvolatile memory modified; see SW2 for details
Execution error	'64xx'	Data in nonvolatile memory not modified; see SW2 for details
	'65xx'	Data in nonvolatile memory modified; see SW2 for details
	'66xx'	Security-relevant result
Checking error	'6700'	Incorrect length (no additional information)
	'68xx'	Functions in class byte not supported; see SW2 for details
	'69xx'	Illegal command; see SW2 for details
	'6Axx'	Incorrect P1/P2 parameters; see SW2 for details
	'6B00'	Incorrect P1 or P2 parameter
	'6Cxx'	Bad L_e value; see SW2 for correct number of available data bytes
	'6D00'	Command code invalid or not supported
	'6E00'	Class not supported
	'6F00'	No specific diagnosis

with the command. At the maximum implementation level, this can involve computing a cryptographic checksum (CCS) or a message authentication code (MAC), a digital signature, or a hash value. In addition, PSO can be used to verify a digital signature or certificate or to encrypt or decrypt data.

The MANAGE SECURITY ENVIRONMENT (MSE) command is used to manage all the security parameters of a smart card. It can be used to configure all the parameters needed for Secure Messaging and the cryptographic functions invoked by the PERFORM SECURITY OPERATION command.

Commands for program code management Smart cards that can download executable program code need several commands to manage the downloaded code. Unlike all other smart commands, the commands for this purpose are not standardized by ISO/IEC, but instead defined by the Open Platform (OP) specification[1] generated by the GlobalPlatform organization.

---

[1] See Global Platform (2003)

The LOAD command loads an executable program into a smart card, after which the INSTALL command informs the operating system that the program is present in the card. The PUT KEY command writes keys to OP applications, while SET STATUS and GET STATUS commands set and read the life cycle status of an OP application. The DELETE command can be used to delete program code, an application or a key that has been loaded into a smart card.

Commands for data transmission   The T=0 transmission protocol does not provide full separation between the transmission layer and the application layer. For this reason, the GET RESPONSE command is provided to enable retrieval of data from a smart card at the application level.

2.3   Data Transmission

Communication with contact smart cards takes place via a half-duplex, bit-serial link. Half-duplex means that only one of the communicating parties can transmit at any given time and the second party receives the transmitted data while the first party is transmitting.

To prevent collisions during data transmission, it is necessary to have a binding agreement that specifies which party initiates communication and a transmission protocol that defines the normal communication processes. In the case of smart cards, the terminal always initiates communications, which means it is the master and the smart card is the slave. The smart card thus transmits data only in response to a request from the terminal.

This master–slave principle pervades all communications with smart cards. Sequence Chart 2.1 on the facing page shows this quite clearly. After the electrical startup of the smart card microcontroller, the terminal sends a reset signal to the smart card, which responds to this signal with an ATR (answer to reset). This can optionally be followed by a PPS (protocol parameter selection), which transfers a set of parameters that modify the subsequent data transmission process. In this case as well, the smart card only responds to an explicit request from the terminal. The actual transmission protocol, during which the smart card only reacts to commands by sending responses, begins after this initialization phase.

There are two types of reset for smart cards: cold reset and warm reset. With a cold reset, the smart card is started up from the power-down state and reset during this process. By contrast, with a warm reset the smart card is already powered up and only receives a reset signal from the terminal.

The communication procedure for contactless smart cards is similar to the procedure for contact smart cards. However, the transmission method used in this case must be able to deal successfully with significantly more difficult communication conditions, owing to the fact that radio-frequency signals are used for data transmission. Abrupt termination of communication due to premature removal of the card from the range of the terminal and transmission collisions due to the presence of more than one card in the range of the terminal are only two examples of difficulties that are relatively common with contactless smart cards. The transmission methods ensure that these and similar conditions can be handled by the communicating parties in a defined manner.

IFD (Terminal)		ICC (Smart Card)
Reset	$\longrightarrow$	Startup smart card operating system
	$\longleftarrow$	ATR
Optional portion		
PPS request	$\longrightarrow$	PPS processing
	$\longleftarrow$	PPS response
Command APDU 1	$\longrightarrow$	Command processing
	$\longleftarrow$	Response APDU 1
. . .		. . .

Sequence Chart 2.1   Logical sequence of transactions during smart card startup. The PPS transaction is optional and can be omitted if the parameters of the transmission protocol provided in the ATR will be used unchanged

The master–slave relationship also affects the behaviour of the chip hardware and the operating system, as illustrated in Figure 2.6. After the power-up sequence, the smart card operating system is started up and an ATR is transmitted. After this, the smart card enters a low-power sleep mode. It remains in this mode until the terminal transmits a command. The command is received and processed, and the response is sent back to the terminal. The smart card then enters the sleep mode again and waits for the next command from the terminal, which causes it to return to the active mode. Alternatively, the terminal can initiate the power-down sequence at this point to shut down the smart card.

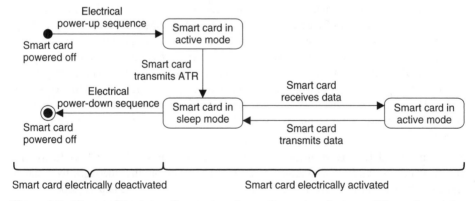

Figure 2.6   The possible states of a smart card operating system for transmitting and receiving data. The smart card remains in the low-power sleep mode until it receives data via the interface. The power-down sequence can be executed at any desired time, but typically, it occurs in sleep mode

2.3.1   Answer to Reset (ATR)

The ATR (answer to reset) is the first communication a smart card sends after detecting a reset. Among other things, the ATR provides the terminal with information about the transmission protocols and data transmission rates supported by the smart card. The ATR is always transmitted with a divider value of 372, which yields a transmission data rate of 9 600 bps with a clock frequency of 3.5712 MHz.

2.3.2 Protocol Parameter Selection (PPS)

PPS (protocol parameter selection) is a communication transaction for changing the parameters used for data transmission between the terminal and the smart card from the values stated in the ATR. These changes are made on the request of the terminal. The PPS is optional and does not have to be executed.

2.3.3 Transmission protocols

The transmission protocols define the communication processes between the terminal and the smart card in case of successful transactions and the mechanisms to be used to handle detected transmission errors.

The most commonly used protocols for chip cards with memory chips are the I^2C protocol and the 2-wire or 3-wire protocol. The T=0 and T=1 transmission protocols, which are commonly used with processor cards, are used almost without exception with contact-type processor cards. There are already several types of smart cards that support the USB protocol, which is widely used in the PC environment. In the case of contactless microcontroller smart cards, the most widely used protocols are ISO/IEC 14 443 Type A and Type B.

Several abbreviations related to data transmission are commonly used in the processor card realm. They are briefly described and illustrated in Figure 2.7 on the facing page. A data record at the transmission level is called a *TPDU* (*transport protocol data unit*), while a data record at the application level is called an *APDU* (*application protocol data unit*). TPDUs and APDUs are defined for the commands sent to smart cards and the associated responses. A command APDU consists of a command header and a command body. The header is mandatory, but the body is optional. A response APDU consists of a response body and a response trailer. Only the trailer is mandatory in the response APDU.

A command APDU consists of four bytes designated as follows: Class (CLA), Instruction (INS), Parameter 1 (P1) and Parameter 2 (P2). The principle that the class byte should indicate the standard in which the command in question is specified is adhered to in most cases. The instruction byte defines the actual command, and the two parameters (P1 and P2) provide additional information about the command.

The command body can contain a maximum of three data elements. The first, L_c (length command), contains the length of the data in the command APDU, while L_e (length expected) contains the length of the data requested from the smart card, which is to be returned in the response APDU.

Four different combinations are permitted for the command APDU. Each combination is called a *case*. There are only two variants for the response APDU. The T=0 or T=1 transmission protocol, which is located below the application layer, looks after communicating these rigidly defined APDUs between the terminal and the smart card.

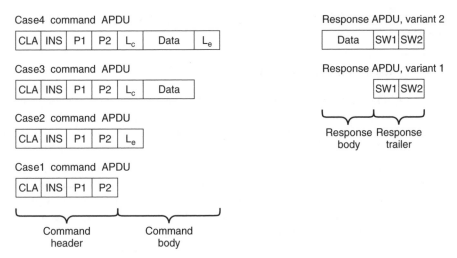

Figure 2.7 The four different cases of command APDUs and the two different variants of response APDUs

2.3.3.1 T=0 transmission protocol for contact cards

The T=0 transmission protocol is the oldest and most widely used protocol for smart cards. It is a byte-oriented transmission protocol with relatively poor layer separation. As a result, Case 4 commands are not possible with T=0. Instead, the terminal must use the GET RESPONSE command to retrieve data to be provided to the terminal by the smart card. However, this has not significantly restricted the use of the T=0 protocol, which is the standard protocol for the world's largest smart card application: the SIMs and USIMs used in GSM and UMTS mobile telecommunication systems.

2.3.3.2 T=1 transmission protocol for contact cards

The block-oriented T=1 protocol has distinct layer separation, so all four cases of command APDUs can be used with this protocol. T=1 has a significantly more complicated structure than T=0, but it is also significantly more robust, thanks to its processes for detecting and resending blocks that contain transmission errors. T=1 is often used with payment cards and ID cards. It is indisputably a more modern protocol than T=0, but its advantages relative to T=0 are not large enough to threaten T=0 with becoming irrelevant.

2.3.3.3 USB transmission protocol for contact cards

The data transmission rate of T=0 or T=1 rarely exceeds 115 kbps in practice. This is too low for smart cards with large data memories. This is one of the reasons why the USB protocol (Universal Serial Bus) is slowly becoming established in the smart card world. The second main reason is that USB provides compatibility with the PC environment. USB smart cards that support the 1.5 Mbps data rate of low-speed USB and even the

12 Mbps data rate of full-speed USB, depending on the microcontroller hardware, are available.

2.3.3.4 Contactless transmission protocols

ISO/IEC 14 443 specifies the properties of contactless smart cards for use at a maximum distance of 10 cm from a terminal. Such cards are called *proximity cards* and they operate on the principle of inductive coupling via an RF magnetic field with a frequency of 13.56 MHz that is generated by the terminal or PCD (proximity coupling device).

Two different transmission techniques can be used for communication, since agreement on a single technique could not be reached during the preparation of the standard. They are called *ISO/IEC 14 443 Type A* and *Type B* and are mutually incompatible. However, commonly used terminals for contactless smart cards, as well as many types of smart card microcontrollers, support both transmission techniques.

The protocol for contactless data transmission between terminals and smart cards uses the half-duplex method and is strongly based on the block-oriented T=1 protocol. There are small differences between Type A and Type B smart cards, but they are relatively insignificant.

2.3.4 Secure Messaging

For some applications, it is necessary to cryptographically secure data transmission to the smart card to prevent eavesdropping and manipulation. This sort of security for smart cards is called *Secure Messaging*. It involves either adding an MAC (message authentication code) to each APDU or fully encrypting each APDU. It is also possible to use send sequence counters (SSCs) for the command and response APDUs to prevent successful playback of previous messages. Secure Messaging is a technically elegant solution that provides transparent communication of APDUs and is highly configurable via parameters, but this comes at the price of complexity.

2.3.5 Logical channels

Some smart card operating systems support multiple logical channels. With multiple logical channels, different applications in a multiapplication smart card can be used fully independently and in parallel during a single session. The logical channel used for communication in each case is addressed using the class byte of the command APDU. This enables the smart card to associate command execution, as well as all attained security and file states, with the selected channel as appropriate.

2.4 Special Operating System Functions

In addition to file management functions, commands and data transmission, smart card operating systems offer a range of special functions that can be used to develop applications. Thanks to hardware support, some of these functions exhibit a level of performance that is certainly comparable to that of modern PCs. The available functions vary depending on the hardware of the selected smart card microcontroller and the operating system, so you should always compare the information provided here against the functional scope of the smart card you intend to use before starting to create a specific application.

2.4.1 Cryptographic functions

The basic cryptographic functions of smart cards encompass the entire range of current cryptographic algorithms. Table 2.5 provides an overview. The basic functions are usually not directly available to the outside world at the interface, but are instead incorporated into commands that provide more abstract functions based on these functions.

Table 2.5   Selected cryptographic algorithms and other algorithms used in typical smart card applications

Type of Algorithm	Algorithm
Symmetric cryptographic algorithms	AES (128-bit, 196-bit, 256-bit) DES (56-bit), TDES (112-bit) IDEA (128-bit)
Asymmetric cryptographic algorithms	DSA ECDSA (160–256-bit) RSA (1024–2048-bit)
Hash algorithms	HMAC MD5 RIPEMD-160 SHA-1 and SHA-256
Key generation for symmetric cryptographic algorithms	Various
Key generation for asymmetric cryptographic algorithms	Various
Random number generators	Various
Error detection codes	CRC and Reed–Solomon

One of these functions is encrypting and decrypting data. This can often be done at the level of performance that is suitable for real-time processing of audio or video data. Another function abstracted from the basic algorithms is authentication of entities, which is usually performed using a symmetric cryptographic algorithm. For compatibility reasons, DES (Data Encryption Standard) and Triple DES are always provided for this purpose, but the trend is clearly heading toward AES (Advanced Encryption Standard) with all three defined key lengths, which is inherently stronger than DES.

In addition to functions based on symmetric cryptographic algorithms, smart cards also offer a full range of signature functions, which are based on asymmetric cryptographic algorithms. The hash algorithms and X.509-compliant certificate management functions associated with digital signatures are also available in smart cards. The actual signature algorithm is either RSA with a key length of 1 024 bits or greater or an elliptic curve cryptographic algorithm (ECDSA – elliptic curve digital signature algorithm) with a key length of 160 bits or greater.

2.4.2 Atomic processes

Atomic processes are an important function for operational reliability of smart card applications. They can be used to ensure that certain data is always written to nonvolatile memory in its entirety or else not at all. If an unforeseen power interruption occurs during an EEPROM write access protected by an atomic process, for example due to the card being yanked out of the terminal, the operating system can restore the original state of the data. However, transaction times are significantly greater when atomic processes are used, since the number of accesses to nonvolatile memory is at least twice as large as with normal processes. Nevertheless, it offers the benefit of reliable protection of important data against corruption due to power interruptions.

2.4.3 Interpreter

For a long time, downloading executable program code to smart cards was taboo for security reasons. However, the benefits of application-specific commands are so large that several different approaches have managed to establish a position in the market.

Downloaded program code is normally interpreted by software in the smart card in order to make it independent of the smart card microcontroller. The program code can be downloaded and executed only after passing suitable security checks, so the virus plague that pervades the PC world is not an issue.

The best-known examples are smart cards that conform to the Java Card specification. However, there are also other operating systems, such as BasicCard and Multos, which provide functionality comparable to that of Java Card. All such smart cards have one feature in common, which is that the operating system provides a large number of interfaces to the interpreted program code. This includes access to the file management system, cryptographic algorithms, other applications in the smart card, and even communication with the terminal. This is necessary and essential to allow interpreted code to be processed at a speed that is acceptable for current applications.

Smart cards of this type are a reasonable choice for all applications in which a specially adapted smart card operating system is not a viable option. The key benefits of this approach – flexibility and independence of the application from the underlying smart card operating system – are responsible for that fact that now, only a few years after its introduction, nearly half of all issued cards have this functionality. Java cards are also used widely in applications with a large number of issued cards, since they make it possible to implement supplementary applications (value-added services) in a SIM or USIM without having to modify the underlying operating system.

2.4.4 Application management

Managing the code-based applications of the card issuer and the applications of independent third parties is a relatively complicated affair. The established industry standard for this is the Open Platform specification.[1] It specifies the procedures for downloading,

---

[1] See Global Platform (2003)

installing, activating, blocking and deleting applications. The associated state diagram is shown in Figure 2.8. The scope of application management is so broad that loading, activating and deleting application-specific keys are also handled by Open Platform functions.

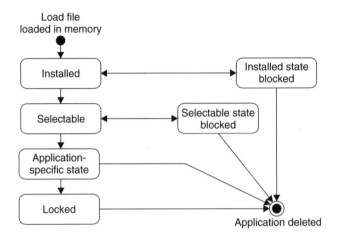

Figure 2.8   States and state transitions in the life cycle of an OP-compliant application. The state transitions are initiated by the Card Manager

The central entity in the Open Platform system is the Card Manager. It runs directly on top of the smart card operating system, which means it acts as the 'deputy' of the card issuer in the smart card. The Card Manager controls the administration of all applications in the smart card. The data needed for this purpose is stored in a sort of database called the *Card Registry*. It contains the most important parameters of the applications administered by the Card Manager, such as the maximum usable memory size. Another entity is the Security Domain. It has a dual function: first as an application-specific storage location for cryptographic keys, and second as the interface to the cryptographic services of the smart card. This provides code-based applications with a set of interfaces they can use to access keys, cryptographic functions, and the administrative functions of Open Platform.

Chapter 3

Application Areas

There is a wide range of potential applications for memory cards and processor cards. The decisive aspect is which parts of the overall application are implemented in the smart card as oncard applications and which parts are implemented in the terminals or higher-level systems as offcard applications. In the extreme case, the smart card contains only a serial number, with the actual application running as a virtual application in the background system. However, this approach falls far short of exploiting the full potential of modern smart cards. It is simply a direct conversion of an outdated system based on magnetic-stripe cards.

The various applications can be classified into closed applications and open applications. In the case of a closed application, the entire system is in the hands of a single entity, such as when smart cards are used as company ID cards. By contrast, open applications involve additional participants that are integrated into the system but do not belong to the system operator. Probably the best representative of such systems is credit cards with chips. In such a system, the issuer is a bank, the system operator is an internationally active credit card company, and the card accepters are local merchants.

3.1 Smart Card Systems

The number of participants in a smart card system can vary over a wide range. In the simplest configuration and in the case of small systems, the participants are often limited to the card issuer and the card users. Larger and more complex systems have additional participants, such as the system operator, application operator, network operator, card accepters, and acquirers. It is certainly not unusual for even more participants to be present in large systems.

Figure 3.1 shows the typical system participants in the case of a small- to medium-sized application. This could be a system using smart cards, for example, as company ID cards, ski lift passes or parking cards.

Smart Card Applications: Design Models for using and programming smart cards   W. Rankl
© 2007 John Wiley & Sons, Ltd

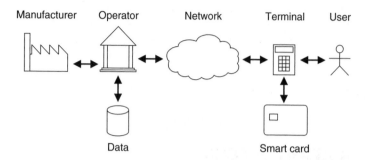

Figure 3.1   General structure of a smart card application with the usual components

In such cases, the operator generates a smart card application (or gets it generated by another party) and obtains its smart cards (including the operating system) from a card manufacturer. Personalization is often performed by the card manufacturer in such cases, but it is certainly conceivable for personalization to be performed by the operator. The operator maintains a server, to which a set of smart card terminals is connected via a network. The smart cards issued to the users are used with the terminals according to the purpose of the application. Nearly all types of smart card systems can be operated using this configuration. If additional participants are necessary, they can be merged into the basic arrangement in the appropriate locations.

3.2   Potential Uses

Smart cards have a wide range of potential applications and can be utilised for many different purposes. However, there are unquestionably areas that are especially suitable for smart cards. The principal characteristics of smart cards are that they can securely store relatively small amounts of data and provide an environment for the secure execution of programs. This makes them excellent candidates for use in the entire security sector. Another important characteristic is that they have a well-established format that is convenient for manual use and handling. Although the smart card interface does not correspond to the current PC and Internet standards, it is still easy to use, which also encourages the use of smart cards.

On the other hand, there are indications that undeniably show that smart cards are not especially well suited to certain applications. One example is any application in which it is necessary to store a large volume of data. Current smart cards typically have a maximum storage capacity of 512 KB. If a data volume in the megabyte or gigabyte range must be managed, other types of media are clearly more suitable. Using smart cards is especially problematic if it is necessary to have special hardware, such as a real-time clock. Modern smart card microcontrollers have a hardware complement that is adequate for current applications, but it will only be extended if the desired production volume is sufficiently large.

A similar situation exists with regard to other possible changes in the specifications, such as a requirement for a temperature range that exceeds the usual range of −25 to

+85 °C (in accordance with TS 102 221). Although such a requirement could be met in theory, the desired production volume would have to be sufficiently large to make this economically attractive for the manufacturer.

In summary, the best possible uses for smart cards are applications that need a large number of inexpensive data storage devices that can securely store individual or personal data and that must perform security-related activities such as authentication, encryption, and/or signing.

3.3 Application Types

Most current PC applications are either file-based or code-based. Websites, which display files in HTML format with the aid of a browser, are typical examples of file-based applications. In the simplest case, no program code is necessary for such applications. The other type of PC application is based on executing a program, such as a word processing program or a spreadsheet program. Such applications also process input and output data, but they additionally require executable software in the target system.

3.3.1 Memory-based applications

Memory cards, which do not contain processors, can be used to implement memory-based applications, which are technically unsophisticated but nevertheless adequate for the intended purpose in many cases. In such applications, the terminal can access the entire memory for read and write operations. Some memory cards require certain conditions to be fulfilled before such access is possible, such as a PIN verification or authentication of the memory card. However, the access logic necessary for this purpose is hardwired in the individual memory chips and cannot be modified. Figure 3.2 shows the architecture of a memory-based smart card application.

Simple applications can be developed quickly using this type of card, but they are limited in terms of their complexity. As memory cards are inexpensive and readily available, memory-based applications are frequently used in relatively small, simple systems. Naturally, this type of application can also be implemented and used with contactless cards.

3.3.2 File-based applications

File-based applications require processor cards and a smart card operating system that runs in these cards. They can be implemented in a single-application smart card or as one of several applications in a multiapplication smart card. A file-based application normally takes the form of a set of data files (EFs) located in a directory file (DF). In addition, the access conditions for reading, searching, writing, creating and deleting the data files are specified by a set of rules. The smart card operating system provides a large number of commands for data access, authentication and other operations. Alternatively, file-based applications can be constructed using data objects (DOs). The latter approach

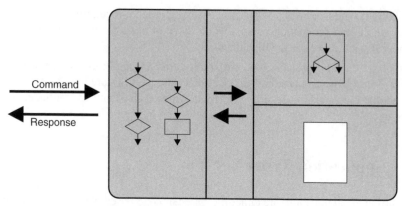

Figure 3.2 Schematic representation of a memory-based smart card application in a memory card. The area with the white background can be modified by the application producer, while the area with the grey background cannot be modified. The commands (which are immutably defined) are shown at the left, the non-modifiable data access conditions are shown at the upper right, and the linearly addressable memory region for user data is shown at the lower right

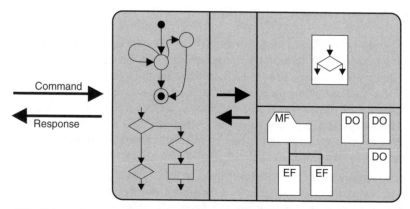

Figure 3.3 Schematic representation of a file-based smart card application in a processor card. The area with the white background can be modified by the application producer, while the area with the grey background cannot be modified. The predefined commands and processes provided by the smart card operating system are shown at the left. The modifiable access conditions for the files and data objects are shown at the upper right, and the various associated data structures for user data are shown at the bottom right

also allows sufficiently granular access conditions to be defined. Figure 3.3 shows the architecture of a file-based smart card application.

These functions for file-based applications are sufficient to implement applications without using program code, even in the case of complex applications. Health insurance cards and signature cards are two well-known examples. They utilise existing operating systems as the basis for corresponding file-based applications.

3.3.3 Code-based applications

Code-based applications also use data files, but the files are complemented by application-specific program code that can be executed in the smart card. This code is usually a Java applet that is managed using Open Platform[1] mechanisms. There are also other options, such as BasicCard, Multos, and smart card operating systems that support downloading of processor-specific program code. Figure 3.4 shows the architecture of a code-based smart card application.

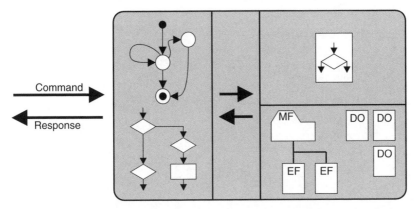

Figure 3.4   Schematic representation of a code-based smart card application in a processor card. The area with the white background can be modified by the application producer, while the area with the grey background cannot be modified. The freely definable program code for commands and processes is shown at the left. The modifiable access conditions for the files and data objects are shown at the upper right, and the various associated data structures for user data are shown at the bottom right

Code-based applications give application developers the largest degree of freedom, since they allow additional commands to be defined to expand the functional scope of the smart card. However, this increased freedom is also a source of errors and incompatibilities. This type of application should only be used if the requirements cannot be fulfilled reliably using a file-based application, and it is only suitable for use by developers who have a certain amount of experience in programming smart cards.

---

[1] See Global Platform (2003)

Chapter 4

Basic Patterns

Before starting to define the actual specifications for a smart card system, you must give some attention to several fundamental issues that are independent of the design and implementation of the system. These issues must be sorted out in the early phases of the project, as they can only be incorporated at a later stage with a disproportionate amount of extra effort and expense. These issues are data protection, export controls, cryptographic regulations, and the underlying standards. When you are starting a smart card project, you must also specify the form and content of the documents to be generated.

If all these items are given proper attention early on, the subsequent phases can be carried out seamlessly and without any complications. The general rule for informatics systems – that the cost of correcting errors made in the early phases increases dispro-portionately with succeeding phases – is especially true in this phase.

4.1 Data Protection

At the nontechnical level and in an abstract sense, smart cards are storage media for personal data. Naturally, this fact arouses considerable interest from a wide variety of governmental and non-governmental organizations.

The first data protection legislation came into force in the 1970s. It included the Fair Credit Reporting Act (1970) in the United States, the Data Protection Act of the German State of Hessen (1970), and the Privacy Act in Sweden (1972). However, data protection did not become established on a broad basis until the 1990s, when nearly all industrial countries passed suitable legislation.

The origins of data protection in Germany are, to a considerable extent, based on the *Volkszählungsurteil* ('census decision') of the Federal Constitutional Court of Germany on 15 December 1983, which formulated the 'right to informational self-determination'. The reasons given for the decision provide an excellent explanation of why data protec-tion is an important civil right:

Smart Card Applications: Design Models for using and programming smart cards   W. Rankl
© 2007 John Wiley & Sons, Ltd

Any person who lacks an adequately reliable general overview of what information about him or her is known in certain areas of his or her social environment, and who is not able to reasonably assess the knowledge of potential communication partners, can be significantly hindered in his or her freedom to make plans and decisions based on his or her self-determination. A social system in which citizens could no longer know what is known about them by whom, at which time and under which circumstances, as well as a legal system that would make such a social system possible, would not be compatible with the right to informational self-determination. A person who is uncertain whether information about nonconforming behaviour is always noted and stored persistently, used or conveyed to others, will attempt to avoid drawing attention by such behaviour.[1]

Adequate protection of personal data should be a matter of course in a modern society. The widely expressed naive argument 'There's nothing to worry about if you have nothing to hide' can be readily contrasted to everyday practice in which anonymity has been taken for granted as long as people can remember. This should also be true in the IT world.

The legal framework has been established by various legislative acts and ordinances, such as the guidelines for personal data published by the United Nations in 1990, the recommendations for protection of personal data generated by the OECD in 1980, the Data Protection Agreement of the European Council in 1981, and in Germany the *Bundesdatenschutzgesetz* (BDSG) (Federal Data Protection Act). This very broad topic, which has now become a special field of jurisprudence, is described below on the basis of general technical principles with particular attention to smart cards.

A high degree of variation prevails among data protection legislation, depending on the particular culture, country, political situation and current events. For instance, in some countries certain data protection measures were largely rescinded after the terrorist attacks on 11 September 2001. Data protection legislation is driven by conflicting priorities, in which it is often necessary to seek a balance of rights between the conflicting interests of the citizens and the state. For this reason, protection of personal data cannot be unconditional because there are critical situations in which society must be able to take action.

Data protection must be achieved by a combination of technology and suitable legislation, although there are unquestionably certain technologies that are 'friendly' to data protection. One example is smart cards, since distributed (noncentralized) data storage in smart cards makes unauthorized access to the data significantly more difficult than if the same data is stored in a central server. If used properly, smart cards can strongly enhance the quality of the protection of personal data.

4.1.1 Definition of terms

Data protection is defined as the technical and legal protection of information regarding personal, professional and business circumstances that can be associated with individual

---

[1] Translation of an unattributed quotation in the German original

persons. Information of this sort is also referred to as personal data. It can be used to identify a person directly or indirectly.

Anonymization means modifying personal data in a manner such that it is no longer possible to associate the modified data with the original person. Pseudonymization, by contrast, means modifying personal data in a manner such that the modified data cannot be associated with the original person unless the rules for associating the data with individual persons are known. The term *pseudonymization* is based on the fact that the simplest way to do this is to replace the original names by unique pseudonyms. The links between the pseudonyms and the original names are made in an association table, which is sometimes called the *association rule*. In this case, deanonymization, which means reversing the anonymization, must not be possible without the association table.

In contrast to data protection, data security means protection against unauthorized access to data or destruction of data.

4.1.2   General principles

Specific data protection measures differ considerably from one country to the next. In the case of systems that are used internationally, finding a compromise that is suitable for all countries where the system is employed can involve major effort and expense, and it can even require imposing restrictions on the functionality of the system. For example, in Germany, compliance with data protection is only obligatory with respect to natural persons, while in other countries such as Switzerland, compliance is also obligatory with respect to legal persons (such as companies).

The basic principles of data protection that are essentially recognized in all areas and countries are listed and briefly explained below. These principles are based on the right to informational self-determination of the involved parties and the basic principles of necessity, use for a defined purpose, and transparency. This conforms to Directive 95/46/EU of the European Parliament and the Council for the Protection of Natural Persons in the Processing of Personal Data.[1] Additional information can be found in the German Federal Data Protection Act,[2] on the Internet site of the German Federal Commissioner for Data Protection,[3] and in a highly comprehensive book by Peter Schaar.[4]

Transparency   Users must always have a clear view of what is done with their data and must be informed of their rights. There must be no concealed actions during read or write transactions at a smart card terminal, and understandable explanations about what is or will be done with the data should be provided.

Use for a defined purpose   The collected data is tied to a particular lawful purpose and cannot be used for some other purpose, not even at some later point in time.

Consent   In the case of data that can be directly or indirectly associated with an individual person, the consent of the person concerned by means of an unambiguous and

---

[1] See EU (1995)
[2] See BDSG (2001)
[3] See BfD
[4] See Schaar (2002)

conscious action is required before the data can be stored or processed. It is furthermore essential that the person concerned is informed clearly and unambiguously of his/her data protection rights.

Necessity   The collected and processed data must actually be necessary for the stated purpose.

Data quality   The processed data must be correct and current.

Data economy and data avoidance   The collected and processed personal data must be limited to the data actually necessary for the purpose and not include any supplementary information. All data that can be used to unambiguously identify a particular person should be stored only as long as absolutely necessary.

Right to information   The persons concerned are entitled to be informed about the processed data at suitable intervals and in an understandable form without unreasonable delay or excessive costs. They are also entitled to correct or delete any data that is incorrect or is contrary to the relevant legislation.

Nondiscrimination   Special protection applies to processing personal data that contains the following information or from which the following information can be derived: racial or ethnic origin, political opinions, religious or philosophical beliefs, union membership, tax status, social status, employment or work status, and data regarding health and sexual activities.

Confidentiality   Personal data collected and stored for a specific purpose cannot be conveyed to third parties.

4.1.3   Recommendations for smart card systems

The following paragraphs are a collection of recommendations for adapting smart card systems currently in common use to comply with the basic requirements of data protection. Additional relevant information and suggestions for general systems can be found in the recommendations of the Data Protection Commissioner of Hamburg,[1] in the article by Hansjürgen Garstka,[2] and of course in the *IT-Grundschutzhandbuch*.[3]

Methods that foster data protection should always be used in smart card systems as much as possible. A typical example is the preferred solution of storing data in distributed form in smart cards under the control of the persons concerned instead of storing data in centralized servers where it is prone to abuse. Furthermore, it must be unambiguously clear to the persons concerned which data about them is processed for what purpose.

You should also not subscribe to the widely shared but mistaken belief that certain personal data is anyhow just trivial data, so the provisions of data protection legislation can be ignored. As a rule, all personal data is subject to data protection and must be processed accordingly.

---

[1] See Datenschutz (1996)
[2] See Garstka (2003)
[3] See ITGH (2004)

Ideally, the system documentation and smart card application will inherently take data protection aspects into account so that suitable checks and controls can be carried out without any problems. The right to informational self-determination also means that the persons concerned can consent to the acquisition, processing and utilization of their personal data by third parties. This is the approach that is usually taken to make it possible to use personal data in a system.

System design considerations   Data protection considerations must be taken into account in the system design phase at the beginning of a smart card project. Data protection is similar to the characteristics of quality and reliability in this regard, since data protection cannot be built in afterward at an acceptable cost. This applies to the entire smart card system as well as the operating system and the actual application in the smart cards. Ideally, a data protection expert should be involved in all the decisions starting from the beginning of the project. It is also helpful to have compliance with the principles of data protection be confirmed by an independent review body after completion of the system architecture phase by means of a data protection audit. Such an audit relatively early in the project makes it fairly unlikely that changes will have to be made at a later stage of the project due to failure to comply with data protection provisions.

Transparency   Transparency of the smart card system relative to the user is an important aspect. Hidden acquisition, reading or storage of data must be avoided without fail. It is important to provide an indication on the terminal that unambiguously informs the user that his or her card is being accessed. Particularly in the case of contactless smart cards, this is often the only way for users to know whether a terminal is communicating with the smart card.

With regard to transparency, it is important to fully inform the user about what is stored in the smart card when the card is issued. This information is often printed in itemized form in the cover letter accompanying the card when it is issued. This fosters a trust relationship between the card issuer and the card user because the user can see exactly which data is stored in the card. Generally speaking, this is handled very well for German health insurance cards. On the other hand, this principle is violated repeatedly by some GSM network operators who modify their service telephone numbers in the background via the air interface without informing the users. Although many subscribers may find this convenient because they always have the current information numbers stored in their SIMs, it creates a certain sense of uneasiness.

Respecting the defined purpose   Acquired and stored personal data is always tied to a specific purpose and cannot be used for any other purpose. This binding of the data to a particular purpose also means that the ever-popular idea (in Germany) of using health insurance cards as 'free' ID cards for companies is fully incompatible with data protection legislation. The data in the health insurance cards is strictly tied to the purpose of health insurance.

Data economy and data avoidance   A fundamental principle is that personal data acquired and stored in a smart card system should be restricted to the data necessary for operating the system. In particular, data logging must leave as few data trails behind as absolutely necessary. In addition, all personal data that is no longer necessary should be deleted immediately and physically. In some cases, it can also be useful to specify a maximum storage period for certain data and delete the data physically at the end of this period. Technical implementation of time-limited storage in smart cards is not easy, as no authentic date and time are present in the card. Nevertheless, the maximum storage period can be made dependent on a transaction counter (to take one possible example) so that the smart card can delete the data concerned independently and automatically when the counter reaches a certain value.

Data access   It should not be possible for data subject to data protection to be read freely from the smart card. If (for example) the log data in a purse card is freely readable, it can be readily used to generate user profiles. To prevent abuse, read access must only be allowed after a successful PIN verification. It should also be ensured, as a matter of principle, that no data intended for other parties can be read, modified or deleted. However, this is difficult to ensure with the freely programmable multifunction terminals currently in common use. The best form of protection has always been and still is to design the smart card application to prevent any sort of unauthorized access to personal data.

GSM mobile telecommunication systems use a clever but nevertheless quite simple mechanism for role-specific data access. It essentially consists of two mutually independent PIN codes, with PIN 1 being assigned to the user of the mobile telephone and PIN 2 being assigned to the owner. The user can make calls with PIN 1, but the data access conditions prevent the user from accessing the data of the owner. The owner can read and write all data after entering the PIN 2 code. This mechanism enables the owner to lend his or her mobile telephone to someone else without allowing the user to read the owner's personal data, such as the last number dialled by the owner. Unfortunately, it is rarely used in practice for reasons of convenience.

Deleting data   If data in the smart card must be deleted, physical deletion should always be given preference because the data in question can be recovered if logical deletion is used.

This principle can be illustrated easily using the SIM cards of the GSM mobile telecommunication system as an example. If an abbreviated dialling number (ADN) is deleted in the SIM, the entire associated record in the EF_{ADN} file must be overwritten with 'FF'. This results in physical deletion of the record. After this, it is no longer possible to read the deleted abbreviated dialling number. By contrast, the TS 51.111 GSM specification states that for an SMS (short message service) message it is sufficient to overwrite the first byte of the record in the EF_{SMS} file with '00'. This byte is a status byte that indicates whether the following data is valid. If this logical deletion is performed, the short message is still present in the EF_{SMS} file and can be read using a READ RECORD command. The reason for using this deletion method, which is questionable from a data protection perspective, is that logical deletion is faster than physical deletion because only one byte has to be changed instead of the entire record.

Centralized storage of data   Centralized storage of personal data is fundamentally problematical because all accesses are outside the control of the users and almost any entity can read or modify the data, depending on the access privileges. Centralized storage should thus be avoided if possible because it is inherently not 'friendly' to data protection and is prone to abuse.

Nevertheless, there are often strong arguments in favour of centralized data storage, since it makes card replacement easy in case of lost or defective cards. It also simplifies the implementation of online data queries by users. Many widely used bonus-point systems operate on this principle, and in many cases they use magnetic-stripe cards or barcode cards solely for user identification. However, other options such as savings stamps, stamp cards or storing data in anonymous smart cards are significantly less questionable with regard to data protection than centralized databases, which are fully opaque to users.

Application partitioning   With modern smart card operating systems, it is a matter of course to fully isolate (partition) different applications present in the same card to prevent each application from accessing the data of another application. However, there are certainly ways to intentionally or unintentionally bypass this partitioning and there are sometimes good operational reasons for doing so. Nevertheless, it must never be possible for an application to access any data belonging to another application that is subject to data protection unless the application accessing the data is authorized to do so.

Anonymity and pseudonymity   Only anonymized data should be used whenever and wherever possible. Deanonymization must not be possible in such cases. This is fully sufficient for statistical system monitoring in most cases. In exceptional cases where anonymity is not possible, pseudonyms should at least be used instead.

Anonymous smart card systems, such as anonymous prepaid smart cards, present significantly fewer operational problems from a data protection perspective than person-specific systems. A good example of pseudonymization can be taken from the GSM mobile telecommunications system. In this system, the IMSI (international mobile subscriber identity) is an internationally unique code that identifies a particular subscriber. However, the IMSI is not used in normal operation. A pseudonym, the TMSI (temporary mobile subscriber identity), is used instead because it is not unique in the overall mobile telecommunication system. Nevertheless, the TMSI is normally adequate for subscriber authentication.[1] This demonstrates that pseudonyms can be adequate for system operation and that it is not always necessary to know the actual identities of system participants.

4.1.4   Summary

Data protection legislation comes into play when personal data is stored or processed. Data protection aspects do not have to be taken into account if the data cannot be unambiguously associated with an individual person. On the other hand, if data protection must be taken into account and storing or processing the data in question is not explicitly permitted by legislation, the person concerned must be informed about the processed data and the purpose for which it is processed. Probably the most effective way to achieve a broadly acceptable form of data protection is to simply refrain as much as possible from collecting and processing personal data.

A major social problem in the current data protection situation is that many people are not aware that they have a personal say in what others do with information about them.

---

[1] See Rankl and Effing (2002)

Although it will take a long time to change this situation, systems in current use should anticipate this change by striving to achieve an adequate level of data protection. Every system operator must always bear in mind that all persons are entitled to protection of their personal data.

4.2   Export Control

Most commodities can be exported to other countries without any restrictions. However, certain categories of commodities are subject to export restrictions or prohibitions based on a variety of international and national statutes and agreements.

At the international level, there is the Wassenaar Arrangement[1] of 1996, which serves as the reference for export controls on armaments and dual-use goods, which includes products that make use of cryptography. Refer to the Wassenaar Arrangement List[2] for details.

Although the export restrictions of the United States do not have any formal validity at the international level, they cannot be ignored in international trade because they can still be applied anywhere in the world under the threat or actual imposition of sanctions.

In Europe, exports are governed by a specific regulation of the European Council[3] and individual legislation in each country. Most countries have similar legislation and restrictions, and in some cases the penalties for unlawful export are draconian. For instance, in Germany even attempting to export restricted or prohibited goods is punishable pursuant to Article 34 Section 5 of the Foreign Trade Act (AWG), and the minimum penalty specified by Article 34 Sections 5 and 6 of the AWG is a prison term of 2 years. Remarkably enough, importing commodities subject to export restrictions is permitted nearly everywhere.

The legal basis for export restrictions and prohibitions in Germany is provided by the *Außenwirtschaftsgesetz* (Foreign Trade Act) (AWG)[4] and the *Außenwirtschaftsverordnung* (Foreign Trade Ordinance) (AWV).[5] The relevant details are specified in the export control list[6] of the Federal Ministry of Economics and Technology, which is regularly revised and can always be revised on short notice depending on the current political situation. This export control list forms the actual basis for export control.

Besides this form of export control, general prohibitions on exports (export embargoes) can also be proclaimed for specific countries and commodities. If an embargo is proclaimed, export of the commodities in question to the country concerned is fundamentally prohibited.

Export restrictions and prohibitions are imposed on commodities in the military, nuclear technology, chemical and biological sectors as well as the information technology sector. This includes devices capable of performing strong encryption and decryption. Smart

---

[1] See Wassenaar Arrangement
[2] See Wassenaar Arrangement: List
[3] See ECR (2000)
[4] See AWG (2004)
[5] See AWV (2001)
[6] See Ausfuhrliste (2004)

cards and security modules are examples of such devices. The restrictions generally apply not only to the export of commodities, but also to the transfer of knowledge to the countries concerned.

In this connection, you must also bear in mind that some commodities can be used for more than one purpose. Such commodities are called *dual-use goods*. These are commodities that can be used for civil as well as military purposes. Depending on their functional scope, it is certainly possible for smart cards to be classified as dual-use goods.

Export restrictions also apply to the essential components of a restricted system, even if these components do not require export licences on their own. This means it is not possible to avoid export restrictions by breaking down equipment that requires an export licence into a large number of component parts that do not require export licences and then exporting them individually.

Similar considerations apply to exporting commodities subject to export restrictions to an intermediate country that does not have any export restrictions for these commodities and then re-exporting them from the intermediate country to the ultimate destination country. Avoiding export restrictions in this manner is of course prohibited. This is legally prevented by the requirement for an end use certificate, in which the recipient in the first country must confirm that the commodities will not be re-exported.

The government agency responsible for export control in Germany is the Federal Office of Economics and Export Control (BAFA).[1] If a commodity requiring an export licence is to be exported, an application for an export licence must be submitted to the BAFA in advance. If the application is approved, the commodity can then be exported. Granting a licence to export a commodity depends on various general conditions. The most important factors are naturally the function of the commodity and the country to which it is intended to be exported.

According to Section 5A002 of the German export control list, all systems, devices, modules and integrated circuits for use in the information security sector that employ cryptographic technology require export licences as a matter of principle. Section 5B002 further restricts the export of equipment for producing the commodities listed in Section 5A002. There is an exemption from this requirement that is particularly important in practice: no export licence is required for devices that only support authentication and digital signatures or symmetric cryptographic algorithms with key lengths less than 56 bits.

Asymmetric cryptographic algorithms whose security is based on factoring integer numbers smaller than 2^{512}, such as the RSA algorithm, are also exempted, along with asymmetric cryptographic algorithms based on the discrete logarithm problem (such as elliptic curves) with an order less than 2^{112}. The list of exempted items could be extended even further, but instead we suggest that you consult the current version of the export control list,[2] which is available free of charge on the BAFA website.

Personal microprocessor cards for mobile telecommunication systems, radio systems, pay TV, digital rights management, copy protection, and payment systems are explicitly excluded from the requirements of Section 5A002 and can thus be exported without a special licence. This means you can take a SIM card in a mobile telephone or an

---

[1] See BAFA
[2] See Ausfuhrliste (2004)

electronic purse card with you to other countries without any problems. However, smart cards that can encrypt and decrypt data using an AES cryptographic algorithm with a 128-bit key length are subject to export restrictions. This also applies to freely programmable smart cards with similar key lengths that permit an unlimited number of accesses to encryption and decryption functions.

4.3   Cryptographic Regulation

After cryptology became a public science in the early 1970s and many cryptographic algorithms and cryptographic protocols were published and became generally available, the question of whether restrictions should be placed on the use of strong cryptography led to a heated political debate in the industrialized Western countries in the 1990s. All cryptographic algorithms that cannot be broken by government bodies are regarded as strong cryptography.

Some thought was given to stipulating that when strong cryptographic algorithms are used, the key must always be deposited with a government body (key escrow). A special cryptographic algorithm called the Digital Signature Standard (DSS), which is only suitable for signatures but not for encryption, was also developed as an alternative. In addition, attempts were made to foster the use of special hardware components (the 'Clipper Chip' and the 'Capstone Chip') in a wide range of equipment, with the associated keys to be held by government bodies in key escrow. Efforts were also made to introduce supplementary key recovery mechanisms in various sorts of cryptographic software. An example is the additional decryption key (ADK) for PGP, which enables companies to access the PGP-encrypted documents of their employees. Another approach is identity-based encryption (IBE). Systems that utilize this form of encryption inherently have an entity that knows all the secret keys of the users. Yet another approach to imposing restrictions as part of government regulation of cryptography was to allow only certain cryptographic algorithms to be used, which if necessary could be broken by government bodies a reasonable amount of effort and expense.

The main argument put forward by the proponents of regulation of cryptographic technology is that it must be possible for government bodies to eavesdrop on communications between persons, in certain cases. The reasons presented by the opponents of cryptographic regulation are more varied. For instance, they argue that strong cryptography is an indispensable prerequisite for using a public network to perform business processes that can withstand a legal audit in case of doubt. In addition, they argue that prohibition would hardly deter criminals from using generally known and readily available strong cryptographic methods for their communications.

Use of strong cryptographic methods would be difficult to prove in practice, since there are several ways to camouflage such use. One way is to first encrypt a message using a strong method and then encrypt the result using a permitted weak method, which is called *superencryption*, while another is to use steganographic techniques to embed a strongly encrypted message in an apparently innocuous message. The consequence of all this would be that government bodies would primarily be able to tap the communications of respectable citizens. Yet another argument is that key escrow and key recovery methods provide definite opportunities for third parties to illicitly obtain secret keys under certain conditions. The report by Hal Abelsen et al.[1] provides a good overview of the issue.

As a result of this debate, in the late 1990s Germany decided to dispense with any legal restrictions on the use of cryptography. This corresponds to the situation in most industrialized western countries. However, there are certainly exceptions to this general rule.[2] In France, encryption was entirely prohibited by law between 1990 and 1996. This severe restriction was somewhat relaxed in 1996. Encryption methods can now be used in France, but it must be possible for government bodies to decrypt the messages if necessary. In Russia, it is still prohibited to encrypt messages without a suitable licence. In the United States, any desired cryptographic algorithm can be used inside the country, but using strong cryptography for communication with other countries can lead to legal problems. It is certainly conceivable, although presently not especially likely, that the current liberal cryptographic policy could be revised under certain conditions.

All cryptographic algorithms can presently be used without any restrictions, and there is no compulsory requirement to incorporate 'back doors' (secret or otherwise) for use by security authorities in systems that use strong cryptography. The absence of such a requirement is highly beneficial to the reliability of such systems and considerably increases confidence in them. For current smart card applications, this means the method that is most suitable from a technical perspective can be used. In this regard, it is by no means always necessary to use the strongest available algorithm, since such algorithms require more processing power and longer keys. On the other hand, excessively weak methods should not be used because that would expose the system to attack by outsiders.

The generally accepted criteria catalogue[3] of the German Federal Network Agency[4] (BNA) can serve as an excellent starting point for your own decisions. For instance, it lists the RIPEMD-160 and SHA-256 hash algorithms as suitable for use until late 2010. In the case of the RSA algorithm, the recommended key length is 2 048 bits if you want to achieve an acceptable level of security over a relatively long term.

4.4   Standards

An important characteristic of smart cards is their broad compatibility with a wide variety of informatics infrastructures. This is largely because of a large number of standards issued by governmental standardization organizations and standards (including industry standards) published by non-governmental organizations. These standards serve as basic reference documents for card manufacturers, operating system developers and application developers.

The international ISO/IEC standards have a very general character with regard to many of their technical aspects. They are thus commonly regarded as a normative framework that forms the basis for additional standards that set out the details for specific

---

[1] See Abelson *et al.* (1997)
[2] See Crypto (2002)
[3] See Criteria (2005)
[4] See BNA

applications. These additional standards in turn form the basis for specifications,[1] which unambiguously set out the actual implementation details.

Availability of the various types of standards has changed distinctly in recent years. In many cases, it is now common practice for standards to be published free of charge on the Internet, and even long-established standardization organizations such as ETSI[2] now make all GSM and UMTS standards publicly available for free downloading via the World Wide Web. Unfortunately, the internationally applicable ISO/IEC standards are not free, but instead must be purchased at relatively high prices as necessary.

4.4.1 Standards for card bodies

The general physical characteristics of cards are described in the ISO/IEC 7810 standard. It forms the basis for a further set of standards (including TS 102 221 and EMV Book 1), which describe specific details and forms of implementation of the ISO/IEC standard in their introductory sections.

4.4.2 Standards for operating systems

The most important set of standards for smart card operating systems is the ISO/IEC 7816 family, which describes the essential informatic aspects of smart cards. The basic data transmission parameters (for ATR, PPS, T=0, and T=1) are defined in Part 3, while the USB protocol for smart cards is described in Part 12. The requirements for contactless data transmission for proximity cards are described in the ISO/IEC 14 443 standard, which consists of four parts.

Part 4 of the ISO/IEC 7816 standard contains a description of the file system, including the file types (MF, DF and EF),[3] file structures (transparent, linear, linear variable, cyclic and TLV-coded),[4] and selection options.[5] The essential mechanisms for Secure Messaging[6] are also specified in this standard. Part 4 of the ISO/IEC 7816 standard is also the most important reference for basic smart card commands.[7] Administrative commands are described in ISO/IEC 7816-9, and commands for cryptographic operations are described in ISO/IEC 7816-8.

Java has become established in the smart card world for applications based on executable program code, and the related specifications published by the Sun Corporation[8] are the most important basic documents for Java in smart cards. There is also a widely accepted standard for loading and managing such code-based applications in smart cards: the Open Platform Specification published by the GlobalPlatform organization.[9]

---

[1] See Section 4.5
[2] See ETSI
[3] See Section 2.1.1
[4] See Section 2.1.3
[5] See Section 2.1.5
[6] See Section 2.3.4
[7] See Section 2.2
[8] See JCVMS (2003), JCRES (2003), JCAPI (2003) and JCAPN (2003)
[9] See Section 2.4.4

4.4.3 Standards for data and data structuring

The ASN.1 method, which is frequently used in smart card systems to structure data, is specified in two detailed standards: ISO/IEC 8824 and ISO/IEC 8825. The primary reference for certificates for public-key infrastructures (PKI) is the ITU X.509 standard. Global data elements and their TLV-compliant designations are described in the ISO/IEC 7816-6 standard.

4.4.4 Standards for computer interfaces

The software aspects of connecting smart card terminals to computers (PCs) are governed by the PC/SC specification, which specifies an interface between smart cards and PC programs. If Java is used as the programming language, the mechanisms of the OCF specification can also be used on top of or in parallel with the PC/SC interface.

4.4.5 Standards for applications

As it is relatively easy to generate applications (either file-based or code-based) using powerful, commercially available smart card operating systems, the standards described below should be consulted and followed according to the specific application type. This is especially advisable because many mistakes commonly made by inexperienced application developers can be avoided entirely by complying with the standards.

Signature applications A wide range of signature cards are now available, but they must be integrated into a wide variety of higher-level systems before their signature functions can be used. At the smart card level, the most important basic document is the ISO/IEC 7816-15 standard, which is based on the PKCS #15 standard generated by RSA. The CEN 14 890 standard, which consists of two parts, governs the essential aspects of signature card design.

Payment system applications Payment system applications are often based on national specifications and are generally tailored to the means of payment and payment infrastructure of a particular country. One of the best-known examples is the German Geldkarte, which is the largest payment system application in the world in terms of the number of issued cards. However, the system specification is not public and the cards are used primarily in Germany. The most important internationally used specifications for credit and debit cards are the four volumes of the EMV specification.[1] Electronic purses in single-application and multiapplication smart cards are described in the EN 1456 standard, which covers nearly the entire range of conceivable variants. This standard is also the starting point for nearly all electronic purse systems currently in use.

Telecommunication applications The GSM mobile telecommunication system is the largest smart card application in the world. The TS 51.100 standard is the primary definitive standard for GSM. It specifies the SIM (subscriber identity module), the transmission protocol, the file system, and the commands. The SIM Application Toolkit,

---

[1] See EMV Book 1 (2004), EMV Book 2 (2004), EMV Book 3 (2004) and EMV Book 4 (2004)

which provides a framework for card-based applications, is specified in the TS 51.014 standard.

New, extensively restructured and expanded versions of the SIM standards were generated for smart cards to be used in the UMTS system. The new standards define a platform called the UICC (universal integrated chip card), which forms the basis for all other applications. The most important application is unquestionably telecommunications using the USIM (universal subscriber identity module). The following standards are also important (along with several others): TS 102.221 for the UICC platform, TS 31.102 for the USIM application, and TS 102.222 for administrative commands. These standards form the essential core set of documents for all smart cards used in telecommunication systems.

All other mobile telecommunication systems, such as the CDMA and TETRA systems, make reference to the GSM and UMTS standards. However, smart cards are usually defined as optional in such systems.

Identification applications   The internationally valid basis for machine-readable travel documents, which typically means ID cards, is specified in the ICAO Doc 9303-3 document. All other details of official ID applications are normally addressed by country-specific standards and/or legislation.

4.5   Documents for Smart Card Systems

A set of documents is necessary to describe a smart card system. The scope of the documents depends strongly on the individual project. For a large national payment system such as the Geldkarte system in Germany, which also has an operating system SECCOS defined specifically for this application, the specifications directly related to the smart card have a scope of more than 1450 pages.[1]

To provide some points of reference, the scopes of the specification documents of several widely smart card systems can be mentioned here. The core specification for the SIM – the smart card used in the GSM mobile telecommunication system – occupies around 180 pages (3GPP TS 51.011 V4.1.0). The specification for the operating system platform of the German health insurance card, which is limited to the requirements level, is approximately 30 pages long, and the actual application specification is approximately 80 pages long. The ICAO specification for machine-readable travel documents[2] is approximately 141 pages long. In the case of credit cards with chips that comply with the EMV specification, the application-independent basic document[3] fills 110 pages and the actual application description[4] fills 164 pages.

---

[1] This includes the following interface specifications for the ZKA smart card: electronic purse application (Geldkarte): 92 pages; electronic cash application: 43 pages; EMV commands: 198 pages; Marktplatz additional application: 39 pages; electronic driving licence additional application: 38 pages; signature application: 304 pages; Secure Chip Card Operating System (SECCOS): 659 pages; concept for personalization of ZKA smart cards with the SECCOS operating system: 77 pages

[2] See ICAO (2002)

[3] See EMV Book 1 (2004)

[4] See EMV Book 3 (2004)

Of course, the specification scope is considerably less for relatively simple applications and applications that use one of the standard smart card operating systems. For instance, if the objective is to develop a smart card system to register flexitime hours or provide access control for a medium-sized company, the smart card application can certainly be described in a specification document of 20–30 pages. This is possible because the other basic documents (such as the description of the transmission protocol or the specification of the smart card operating system used for the application) can be included by reference.

The documents of a smart card system normally have a clearly defined hierarchy. At the highest level of abstraction, the customer requirements specification contains a description of the initial situation and the system requirements. These requirements are independent of the implementation and define *what* the new system is to do. The customer requirements specification is normally generated by the contracting authority, and it forms the basis for the bidder's proposal and cost estimate.

The next level in the hierarchy is occupied by the specification document generated by the contractor in response to the customer requirements specification, which goes by various names such as engineering specification or product specification. It presents the contractor's view of *how* the requirements stated in the customer requirements specification will be fulfilled. At this point, the level of abstraction is still high enough that this specification can be understood by contractors having a relatively low degree of technical sophistication.

The detailed specification documents are located at a level of abstraction directly above the source code. Their purpose is to describe the technical components of the smart card system unambiguously and in a manner that is not open to interpretation.[1]

Subjunctive terms such as 'could', 'should' or 'ought to' and phrases such as 'would be desirable' or 'is to be considered' have no place in specifications. Otherwise, the developers will not have unambiguous specifications, with the result that they will consciously or unconsciously choose one of the possible options according to individual circumstances. In the worst case, the development team responsible for some other component will take a different approach and in the integration phase the integration team will discover that the two components are incompatible – for example, smart cards and the terminals may not work together properly for certain functions.

A comment regarding the literary ambitions of many specification writers is in order here. The purpose of a specification is not to demonstrate that its author is an expert on a particular subject, but instead that the author knows how to describe technical facts and conditions clearly and unambiguously using simple wording. There are specifications in which the DES cryptographic algorithm is described in great detail over many pages. This is simply a wasted effort that can even lead to misunderstanding if the implementer assumes it describes a variant of DES, and it distracts attention from the essential content. In such a situation, it would be fully adequate to specify 'an SPA/DPA-resistant, noise-free DES algorithm with an execution time less than 2 ms per 8-byte block' instead of generating several pages of process descriptions.

---

[1] Unfortunately, there is not enough room here to describe all the steps of the process that leads the requirements to the detailed specification. The description of the V model (V-Model XT, 2004) provides a very extensive and exact explanation

In practice, the customer requirements specification and the engineering specification are sometimes combined into a single document in the case of relatively small projects or when the customer has a good understanding of the subject.

The hierarchical chain from the ultimate requirements to the detailed specification can be illustrated using a simple example. A typical requirement might be that the background system must be able to verify the authenticity of the smart cards. In response to this requirement, the engineering specification might state that authenticity of the smart cards is to be verified using a challenge–response transaction based on the AES algorithm. The detailed specification, which is generated based on the content of the engineering specification, might then state that authentication of smart cards by the background system is to be performed using the INTERNAL AUTHENTICATE command as defined in ISO/IEC 7816-4 with the AES cryptographic algorithm and a card-specific key with a length of 256 bits.

4.5.1 Specification partitioning

In nearly all cases, the documentation of a smart card system can be partitioned into a specification document for the overall system, a specification document for the background system, a specification document for the smart cards, and a specification document for the terminals. The scope and emphasis of the individual specification documents can vary considerably depending on the specific nature of the system.

4.5.1.1 System specification

The system specification addresses all high-level aspects of the smart card system. These aspects relate to operational use and the interactions of the individual components.

System overview   This is a general overview of the entire system. The smart cards are only a small part of the entire system. The system overview explains the relationships between the components and describes the basic characteristics of the components. This document can also be regarded as a summary for the 'reader in a hurry'.

Security architecture   This specification can contain descriptions of key hierarchies, key generation, key derivation, key referencing, key exchange, the cryptographic algorithms used in the system for the functions of encryption, signing, hashing and generating random numbers, and the cryptographic protocols used in the system. It may also contain security requirements and a list of measures against potential attacks. This document is often confidential and only available to persons who need it to perform their regular work.

Data dictionary   This compilation, which is sometimes also called a *data description directory*, consists of a list of all data objects used in the system and their coding.[1]

4.5.1.2 Background system specification

The specification for the background system describe all the functions at the level or levels above the smart card terminals. This includes system operation, system monitoring,

---
[1] See Section 5.1

card management, billing or settlement, and interfaces to other systems. The documents forming this specification can be limited to a few pages for simple applications, but they can easily amount to several thousand pages in the case of highly complicated systems such as the GSM mobile telecommunications system.

4.5.1.3 Smart card specification

The smart card specification consists of the specification documents for the card body, the operating system, the application as seen from the user perspective, special applications, and personalization. The informatics aspects of this specification, which is the key document for small- and medium-sized applications, are described in Section 4.5.2.

Card body   The card body is usually more than just a medium to house the chip module – it is also a design element and an advertising medium. This means that specifications are also necessary for elaborate colour printing, transparent features, holograms and the like. The sizes and positions of many card elements,[1] such as signature panels, magnetic stripes, embossing and the chip module, are specified in relevant standards such as ISO/IEC 7810, ISO/IEC 7811, and ISO/IEC 7813.[2] Compliance with these standards is important, as otherwise there is a significant risk of problems when the smart cards are inserted into terminals.

Smart card operating system   The operating system, including its transmission protocols, file management functions, commands and state machines, is described in this specification. If downloadable program code is supported, the specification also describes the supported software interfaces and the mechanism for loading program code.

Main application   This specification defines the details of the main end-user application built on top of the smart card operating system, including its data, commands and processes.

Additional applications   Applications that can be included in the smart card in addition to the main application are specified individually in separate documents.

Hardware security module   If the smart cards are used as hardware security modules (HSMs) in certain types of equipment (such as terminals), the associated application is described in this specification.

Personalization   Initialization and personalization[3] require special commands and processes that can only be used in a particular phase of the life cycle and are disabled in normal use. For this reason, the informatics aspects of card production are specified in a separate document.

Migration concept   This document, which is only generated in isolated cases, describes the detailed transition process for changing from an existing card generation to a new card generation with a different technical configuration.[4] All related modifications to or

---

[1] See Section 1.3
[2] See ISO/IEC 7810 (2003), ISO/IEC 7811-1 (2002), ISO/IEC 7811-2 (2001), and ISO/IEC 7813 (2001)
[3] See Section 7.1
[4] See Section 7.2

adaptations of all components of the system, including the background system, personalization and the terminals, must be addressed. The objective is to ensure that both card generations are fully functional during the transition period.

4.5.1.4 Terminal specification

The scope of the specification documentation for the terminals used in the system is highly dependent on the nature of the system. For example, payment systems have separate specification documents for merchant terminals, vending machine terminals, loading terminals, special-function terminals, administrative terminals, and pocket terminals. Simple systems often have only one type of terminal for all transactions with card users. Some systems also have special administrative terminals so that authorized persons can modify the data in the smart cards.

4.5.2 Elements of a typical card specification

Modern specification documentation in the smart card arena is more or less fully partitioned into specification documents for smart cards, smart card operating systems, and smart card applications. Particularly in the case of large, complex systems, this has the advantage that the basic components only have to be described once, after which they can be included in the various application specifications by reference. This situation is reflected in the examples of specification documents described below. However, in the case of relatively small applications it is by no means unusual to generate a single comprehensive document with three main sections instead of three separate documents.

4.5.2.1 General information

A general information section should be included in every specification document. The purpose of this section is to provide a reference framework for the reader and an introduction to the specification document.

Introduction Every specification document should have an introduction that describes the scope of the specification. The introduction should also include a list of abbreviations and notation used in the document. A list of the standards referenced in the document (normative references), a bibliography, and definitions of the most important terms used in the document also belong to this part of the document.

Change history If specification documents are distributed to an extended group of persons or organisations, it is essential to document all changes made to the documents after the initial distribution. In the simplest case, a paragraph listing the major changes and dates can be added at the end of the document. However, in many cases it is highly advisable to also insert change bars in the page margins when a new version is issued. In large documents, this is the simplest and most reliable way to clearly indicate changes to the reader.

Appendix Examples and specific instances of the situations described in the body of the document are often included in an appendix. Particularly in the case of complex data coding, nested ASN.1-structured data objects and certificates, such examples are quite valuable and sometimes indispensable for proper understanding of the document. Some specification documents also include detailed descriptions of typical command sequences in order to clarify common use cases. Incidentally, these command sequences should always be incorporated in the related tests.

As the patent world (or madness) is playing an increasingly prominent role in the informatics area, it is also a good idea to include a list of patents or intellectual property rights (IPR) related to the specification.

The appendix is essentially a catch-all for important information that does not fit into the body of the document but nevertheless deserves mention.

4.5.2.2 Smart card

All general, physical and electrical parameters of the smart card are set out in the smart card section of the specification document. The informatics aspects are dealt with in later sections. This section is unnecessary if the application uses an existing type of smart card.

Physical characteristics The physical characteristics section describes the physical properties of the card body and the chip. The most important aspect here is the size of the card, which is usually the ID-1 format.[1] The temperature range, which must also be stated here, is also an important parameter because it determines which types of plastic can be used for the card body.

The contacts, which form the electrical interface to the microcontroller, are also included in the specification of the physical characteristics. This includes the dimensions and locations of the contacts and the maximum permissible contact pressure. The active contacts must also be specified.

Electrical characteristics The most important part of this section is the description of the electrical characteristics of the standard contacts: supply voltage (Vcc), ground (GND), data transmission (I/O), clock (CLK), and reset (RST).

This section also includes a specification of the maximum current consumption of the microcontroller as a function of the supply voltage, which is an especially critical parameter for cards used with mobile terminals. The minimum and maximum clock rates and specification of whether stopping the clock is permitted are important criteria for the clock source. The power-up and power-down sequences of the microcontroller are also described here.

As the electrical characteristics of commonly used smart card microcontrollers are normally based on the ISO/IEC 7816-3 or ETSI TS 102.221 standard, these standards are usually included by reference for applications with a small- to medium-sized number of smart cards.

4.5.2.3 Smart card operating system

The smart card operating system specification is usually a document that describes the behaviour of the interface between the smart card and the terminal. The actual

---

[1] See Section 1.2

implementation of the operating system is not addressed by the specification document, for the simple reason that an interface description is significantly less complex than a full specification of a complete operating system.

Data transmission   Data transmission specifications are usually quite extensive. However, the scope can be reduced markedly by incorporating a suitable specification document by reference. In such case, you should be wary of referencing overly generic standards (such as ISO/IEC 7816-3), since they provide such a large number of options that no smart card operating system supports all the available options.

This document always begins with a description of the ATR[1] (Answer to Reset) and its data elements, as well as the optional PPS[2] (Protocol Parameter Selection) sequence for changing to a different transmission protocol or different protocol parameter values.

In case of a new application, it is sufficient to select one of the two commonly used transmission protocols: T=0[3] or T=1.[4] The more recent, block-oriented T=1 protocol is generally preferable to the older, character-oriented T=0 protocol. In practice, the only reason for supporting both protocols is for compatibility. A reference to the appropriate standard for the selected protocol should be stated, along with any restrictions as appropriate if only some of the functions of the protocol are to be supported.

Some relatively recent smart cards also support the low-speed (1.5 Mbit/s) and full-speed (12 Mbit/s) versions of the USB protocol. It is sometimes possible to use the USB protocol in parallel with T=0 or T=1. This must be described in detail in the operating system specification. As USB has been relatively rare in smart card systems up to now, the related documents should be sufficiently detailed and all the characteristics of the USB protocol should be specified.

Commands   The command definitions are normally limited to descriptions of the input and output data and a textual description of the actual function of each command. This level of detail is sufficient for nearly all specification documents. However, this leaves the order of the return codes undefined in case of multiple errors in the input data. This is rarely significant in practice because the terminal usually terminates the transaction if an unexpected return code is received. If it is necessary to specify the exact sequence of the requests and return codes, the operation of the command must be described in pseudocode.

Table 4.1 illustrates the usual manner of describing a smart card command, Table 4.2 illustrates the associate response to the command, and Table 4.3 shows the possible return codes.

4.5.2.4 Application

An application in a smart card essentially consists of a set of files or data objects and the commands that use this data. All of these elements must be exactly specified to enable the application to be implemented in the smart card.

---

[1] See Section 2.3.1
[2] See Section 2.3.2
[3] See Section 2.3.3.1
[4] See Section 2.3.3.2

Table 4.1 Description of the READ BINARY command as an example of a commonly used structure for describing smart card commands

Data Element	Length	Content	Description
CLA	1 byte	'00'	Class byte as specified in ISO/IEC 7816-4, without using Secure Messaging
INS	1 byte	'B0'	Instruction byte for READ BINARY as specified in ISO/IEC 7816-4
P1	1 byte	...	Parameter 1 P1.b8 = 0: Read data from the currently selected file with an offset; offset = (P1.b7 ... P1.b1 ‖ P2) P1.b8 = 1: Read data after implicit file selection using a short FID with offset; P1 = (100 ‖ short FID); short FID = (P1.b5 ... P1.b1); offset = P2
P2	1 byte	...	See description of P1
L_e	1 byte	...	Expected length L_e = 0: Read all data until the end of the file L_e > 0: L_e is the number of data bytes to be read

Table 4.2 Description of the response to the READ BINARY command as an example of a commonly used structure for describing smart card commands

Data Element	Length	Content	Description
DATA	n bytes	...	Data read from the file, with length n L_e = 0: n = file length L_e > 0: $n = L_e$
SW1	1 byte	'90'	Status words 1 and 2 if no error occurred
SW2	1 byte	'00'	See description of SW1

Table 4.3 Description of the return codes of the READ BINARY command as an example of a commonly used structure for describing the return codes of a smart card command

SW1	SW2	Meaning (All return codes in accordance with ISO/IEC 7816-4)
'62'	'81'	The returned data may contain errors
'62'	'82'	Fewer than L_e bytes could be returned because end of file was reached first
'65'	'81'	A memory error occurred
'67'	'00'	Incorrect L_e parameter
'69'	'81'	Command incompatible with file structure
'69'	'82'	Required security state for read access not satisfied
'69'	'86'	Command not allowed because no file selected

Data elements   The most common way to manage data elements is to use a data dictionary, ideally in the form of a database. A data dictionary is not essential for relatively small applications, but even then it is often quite useful as a general reference for the data elements, since the files and data objects can be specified based on the data dictionary. The structure of a data dictionary is described in detail in Section 5.1, including several examples.

Files and data objects   Files and data objects are composed of data elements. The relationships between the data elements and the files and data objects are defined in the application specification.

The positions of the files in the file structure (file tree) of the smart card must be agreed on, suitable file names (FIDs – file identifiers) and short file identifiers (SFIs) must be specified for the data files (EFs – elementary files), and a suitable file structure (including its size parameters) must be specified. Each directory file (DF – dedicated file) requires a DF name for identification, which may contain an AID (application identifier). It is also important to specify the access conditions for the data files. Table 5.5 shows the structure of a log file and its associated attributes as an example of a file structure.

Data objects in smart cards can be identified using specific tags (ID codes). These tags are sometimes included in a template that is administered by the smart card operating system. For reliable referencing, it is usually necessary to assign the data objects to a DF, and it is also necessary to specify the access privileges for each data object.

Commands   The commands necessary for an application can either be selected from the predefined commands of the smart card operating system or implemented in the form of executable program code loaded into the smart card. If operating system commands are used, it is sufficient to include a reference to the appropriate document in the specification. Special commands realized using application-specific program code must be described unambiguously and in detail so that they can be implemented correctly. It is a good idea to model such descriptions on the specification of the smart card operating system as described in Section 4.5.2.3.

If mechanisms for Secure Messaging[1] or logical channels[2] are used, they must be specified.

Application protocol   The command sequences to be used for the application are described in the application protocol using typical scenarios. An example would be the sequence of all commands necessary for full mutual authentication of a smart card and a terminal. The most commonly used scenarios are selecting an application, identifying a person, authenticating a device, reading and writing specific data, blocking and unblocking files, creating and deleting files and applications, downloading programs, and signing and verifying data.

4.5.3   Document distribution

It is an undisputed principle of security systems that the security of the system must depend solely on maintaining the secrecy of the keys used in the system. This is also

---

[1] See Section 2.3.4
[2] See Section 2.3.5

called *Kerckhoff's principle*. The consequence is that with the exception of secret keys, everything associated with such a system can be made public without compromising the security of the system. Many major smart card systems comply with this principle to a large degree. Some examples are the GSM and UMTS mobile telecommunication systems, debit and credit card systems that comply with the EMV standard, and the German Geldkarte system (a national payment card system).

Nevertheless, you should bear in mind that making all the documentation of a system public gives potential attackers access to a wealth of information that would otherwise cost them considerable effort and expense to obtain. Consequently, a mixture of publication and secrecy is often used in practice. The majority of the specifications are usually public and can thus be reviewed by experts without any special considerations regarding confidentiality, while a relatively small portion of the specifications, which are specifically relevant to security, are confidential. For example, many GSM network operators keep the cryptographic algorithms they use secret. However, the cryptographic algorithms are analyzed in detail by various technical experts before they are used, in order to assess their strength. Suitable personal confidentiality agreements ensure that no information about the results of the assessment is made public.

An important consideration is that keeping documents secret should only be used as a means to create extra effort and expense for potential attackers, rather than (for example) to camouflage the security weaknesses of the system. There are many examples that provide compelling evidence that keeping specifications secret does not provide any genuine protection against the discovery of security weaknesses of systems in widespread use.

For instance, keys with insufficient length were used for many years in the French Yescard payment system, with the result that an ambitious attacker found it relatively easy to bypass the entire security system in 2000.[1] A quite similar situation existed with an RFID-based automotive anti-start system made by Texas Instruments, which was used by many car manufacturers. The security of this system was broken in 2005 due to the fact that the key was too short.[2] The specifications of both these systems were confidential, which made the jobs of the attackers considerably more difficult, but ultimately this was not enough to ensure the security of the systems.

4.5.4 Document version numbering

Numbering the versions of a document appears to be rather simple at first glance: you simply use an integer that is incremented each time a new version is generated. It is also reasonable to supplement the version number with a date to make the time history clear. The initial version number is usually 1, although 0 is sometimes used to identify an unstable preliminary version. This scheme is well suited to documents that are not subject to overly frequent changes. However, the version numbers can become quite large rather quickly if the documents are revised frequently, regardless of whether the changes involve correcting spelling errors or rewriting half the document.

A version numbering system based on three independent numbers separated by full stops was introduced some time ago by ETSI for the standards of the GSM mobile

---

[1] See CLUSIF (2002)

[2] See Bono *et al.* (2005)

telecommunication system. The first number is increased when major changes to the content of the document are made. The middle number is incremented for minor additions, clarifications, technical corrections and revisions. Layout changes, spelling corrections and similar revisions that do not affect the technical content cause the last number to be incremented. This provides readers with a quick overview of the changes made to each document.

For example, if the version number of a document is increased from 5.3.1 to 5.3.2, it is immediately apparent that no significant changes have been made to the content of the document. By contrast, a document whose version number has been changed from 5.3.1 to 6.0.0 has experienced major changes to its content. This scheme is quite suitable for documents subject to constant revision and upgrading.

Unfortunately, this numbering scheme was not maintained when the specification documents for UMTS were generated as part of the 3GPP project in the late 1990s. Instead, a reference to the year number was introduced and the specification numbering scheme, which had been used for more than a decade, was completely rearranged.

Numbering schemes such as the system used for the text composition program TeX, which leads to version numbers that approach the value of π, should be avoided because they become difficult to comprehend. The three-figure numbering system used for the Linux kernel, in which an odd-valued middle figure (such as '2.3.1') indicates a developer version and an even-valued middle figure (such as '2.4.1') indicates a stable version, is also not particularly suitable. All version numbers driven by sales and marketing considerations, which are usually maintained for only a short time and are quickly sacrificed to achieve the newest surprise effect, are equally unsuitable.

A single-figure version number scheme is advisable for documents that are modified only rarely, while a scheme using three separate figures, similar to the ETSI scheme, is advisable for documents that are revised frequently. The date should also be stated to make it easy to grasp the time history.

Chapter 5

Architecture Patterns

An architecture consists of designs and descriptions that define the properties of the object to be produced. The activities and tasks in the architectural phase are abstract and still far removed from the implementation level. However, design models and prototypes are highly beneficial in this phase and they can be used quite effectively. This chapter is thus fairly extensive in order to address the most important architectural aspects of smart card applications.

5.1 Data

It is decisively important to always maintain a good overview of the data in a smart card system. One way to do this is to generate and maintain a data dictionary of all the data in all components of the system. In the simplest case, this dictionary can consist of a table generated using a word processing program, but relatively complex database applications are often used for this purpose in complex systems. Table 5.1 shows an example of a typical entry in a data dictionary.

All the data elements used in the system and their essential attributes are listed in the data dictionary. A name that is unique in the system should always be present as a key entry, which can be used as a reference in the entire system.

An example is the ICCID (integrated chip card identifier), which is a data element of the SIM used in the GSM system. It is the unique identifier of a particular smart card in the system during the entire life cycle. Some other examples of data element names that are unique in the entire system are KDEnc1 (key derived encryption no. 1) and IMSI (international mobile subscriber identity), which is an internationally unique identification code for a mobile telecommunication system subscriber. It is also a good idea to briefly describe the purpose of each data element so that the reason for the existence of the data element can be discovered quickly without ploughing through specification documents. The size of each data element, and especially its coding, are important for proper cooperation between the components of a smart card system. As

Smart Card Applications: Design Models for using and programming smart cards   W. Rankl
© 2007 John Wiley & Sons, Ltd

Table 5.1 Some typical data elements of the data dictionary of a smart card application

Description	Example 1	Example 2
Data element	PIN	PIN_EC
Description and use	PIN (personal identification number) of the card user	PIN error counter
Size and value range	8 bytes	1 digit [0 ... 9]
Format	Bytewise ASCII coded, left aligned and padded with blank characters (ASCII '20')	
Template	–	–
Tag	'99'	'9F17'
Storage location	MF.EF$_{PIN}$	MF.EF$_{PIN}$
Read by	Never; only read internally by the operating system with value returned after PIN verification	Value returned after PIN verification
Modified by	Can be modified after correct PIN entry	Managed by smart card operating system
Initial value	Random value generated during personalization	3, configured during initialization
Example (logical)	1234	–
Example (coded)	'31 32 33 34 20 20 20 20'	–

these codings are frequently the source of misunderstandings, each coding should also be illustrated by at least one relevant example.

The individual data items are normally entered in the smart card during personalization. It is thus useful to state the initial values of the data elements in the data dictionary as well as the entity that generates the data or acts as the source of the data. Stating where the data is stored in the smart card is also useful for the same reason, as this information is used during personalization and subsequent operational use. To ensure that the correct access conditions are configured when the data is stored in the smart card, the data dictionary should also state who can read and modify the data concerned.

Ideally, you should generate and maintain a data dictionary starting with the specification phase of the smart card project. A large number of changes will necessarily occur during the entire development process until the system is launched and put into operation. It is therefore advisable to place the data dictionary under version management. Ideally, the change history should extend down to the data element level.

5.2 Data Coding

Appropriate coding of the data used in the smart cards and smart card system saves memory and reduces the number of conversions that have to be performed during data processing. The relevant data structuring standards are described in Section 4.4.3, and the most important principles for data used in applications are described in Section 6.1.3. It is important to code all data to minimize the use of memory space, avoid any form of redundancy, and restrict the data stored in the smart card to what is actually necessary. On the other hand, you should avoid overzealously stripping down the data to a point that makes future expansions or upgrades impossible. This means that you should weigh your decisions judiciously and look beyond the immediate situation.

Coding data for smart cards does not involve anything especially unusual. Numbers are almost always limited to unsigned positive integers, including zero. Floating-point numbers, which are often used in the PC world, are not used in smart cards – not even in electronic purses. The disadvantage of floating-point numbers is that they are complicated to use even with basic operations and they also lead to the well-known rounding problem, which is undesirable in payment systems. Electronic purse applications in smart cards thus use only integers to represent monetary amounts, with the basic unit corresponding to the smallest monetary unit (such as 1 cent). This means that fractions of the basic unit cannot be represented, but that is insignificant in normal practice.

A value range of one or two bytes without additional coding is usually used for numerical values that are further processed by terminals or other IT equipment. However, if integers are displayed in decimal form in the user domain, BCD (binary coded decimal) coding is the prevailing scheme. Parameters associated with Yes/No decisions are coded using bit fields in the smart card domain, since they occupy relatively little memory space and can be represented in an easily understandable manner.

The situation regarding character coding is just as muddled in smart card systems as in the PC world. The 7-bit ASCII (American Standard Code for Information Interchange) code is often used, but the set of characters that can be represented using this code is insufficient for the vast majority of users in the world. Consequently, the UCS-2 (universal character set) representation has become established in recent years, even though it requires 16 bits for each character. UCS-2 is also compatible with the widely used Unicode system. It enables the most important characters of all cultural groups to be represented, and the additional memory requirement is acceptable in most cases.

5.3 Files

The vast majority of smart card applications are file-based applications[1] consisting of a certain number of files (EFs – elementary files) with corresponding access conditions, all located in a directory (DF – dedicated file). The most important task for generating an application thus consists of specifying the files and associated access conditions.

Ideally, you should start with a data dictionary[2] containing a list of all the data elements used in the application. The next step is to organize the data elements on the basis of how they are used and the read and write privileges defined for the administrative and operational phases. Naturally, these access conditions only apply to accesses from

---

[1] See Section 3.3.2
[2] See Section 5.1

Table 5.2   Data elements for a typical access control card and the associated read and write
conditions for the administrative and operational phases. 'ADM1' is the administration
PIN for card personalization and 'ADM2' is the administration PIN of the system
operator

| Life cycle Phase | Administration Phase | | Operational Phase | |
Access	Read	Write	Read	Write
Card number	Always	ADM1	Always	Never
Seed for PRN generator	Never	ADM1	Never	Never
Current random number	Never	ADM1	Always	Never
PIN	Never	ADM1	Never	PIN or PUK
PIN error counter	Always	ADM1	Always	Never
Access privileges	Never	ADM1	PIN	ADM2
Access protocol	Never	ADM1	PIN	Never

outside the smart card, but not to internal access by the smart card operating system.
Table 5.2 shows an example of data elements organized in this manner.

5.3.1   Access conditions

The next step is to define a systematic set of access conditions (access privileges). These
conditions essentially relate to user identification using PIN verification and unilateral or
mutual authentication of the smart card and/or components of the outside world. Enter
the results in the previously mentioned list of data elements. Next, you can systematically
group the individual data elements into separate files, which form the basis for the file-
based application. As part of this activity, you can also specify the file structures of
the individual files. Table 5.3 shows some of the data elements of Table 5.2 assigned to
several files.

A few errors are bound to creep into the file definitions, even if you follow the above
steps carefully, so you should always use all typical use cases to check the formulated
list of files and access conditions for reasonableness. Correct the file list based on the
errors you discover during this check.

The next step depends on whether the smart card operating system you intended to use
supports state-based or rule-based access conditions.[1] If it supports state-based access
conditions, the access conditions in the file list must be aligned to the states provided
by the operating system. This may well require making a few compromises in your
original, ideal scheme. However, more than a decade of experience with smart card
applications has shown that even very large smart card applications, such as the SIMs
used in the GSM system, can be generated in a practical and future-proof manner using
such relatively simple access conditions.

Rule-based access conditions provide significantly more freedom for specifying file access privileges. However, this also creates significantly more complexity and thus more

---

[1] See Section 2.1.6

Table 5.3 Assignment of some of the data elements listed in Table 5.2 to files according to the specified read and write privileges

Data Element	File	EF$_{ARR}$ Rule Listed in Table 5.4
Card number	EF$_{Cardnumber}$	SE1, Rule set 2
		SE2, Rule set 1
Random number generator seed	EF$_{RNDSeed}$	SE1, Rule set 3
		SE2, Rule set 4
PIN	EF$_{PIN}$	SE1, Rule set 3
		SE2, Rule set 2
Access privileges	EF$_{Privl}$	SE1, Rule set 3
		SE2, Rule set 3
Access protocol	EF$_{Prot}$	SE1, Rule set 3
		SE2, Rule set 5

opportunities to make mistakes. A prerequisite for this type of access control is an EF$_{ARR}$ (access rule reference) file in the DF of the application. Each record of the EF$_{ARR}$ file contains a set of rules for accessing a particular file. Table 5.4 shows some typical access rules that could be placed in an EF$_{ARR}$ file for the data elements and files listed in Table 5.2 and Table 5.3.

You must generate a set of rules corresponding to the previously generated list of files and associated access conditions and then distribute them among the appropriate records of the EF$_{ARR}$ file. In the interest of simplicity, it is appropriate to note here that you should be economical when generating access rules. Each opportunity to omit a rule by combining two or more rules yields a significant reduction in complexity. As a guideline, we can point out here that even complicated applications rarely require more than ten rules in the EF$_{ARR}$ file.

All entities involved in the entire life cycle of the smart card must be taken into account when defining access privilege groups. Initialization and personalization by the card manufacturer occur at the beginning of the smart card life cycle. They are followed by an administrative phase with an application operator. It is certainly possible for several applications belonging to different operators to be present in a single smart card. This must be reflected in the access rules. Specific privileges are usually necessary for the smart card user, and possibly also for the card owner, although these privileges can often be combined.

The access conditions for the EF$_{ARR}$ file must be chosen carefully because this file governs all accesses to the files of the smart card application. If the rules in EF$_{ARR}$ can be modified, the entire security scheme can be bypassed. Consequently, write accesses to EF$_{ARR}$ must be restricted to the administrative level and users must never be granted write access. If it is possible to foresee that the specified access rules will be adequate for any files to be created at some later date, write access to EF$_{ARR}$ can also be set to 'Never'.

Of course, it must be impossible to delete EF$_{ARR}$, as otherwise the EF$_{ARR}$ file at the next higher level would become applicable and the access rules defined in that file could lead to security problems.

Table 5.4 Example of the typical content of an EF_{ARR} file for a system with two different security environments (SEs): one for the administrative phase (SE1) and the other for the operational phase (SE2)

SE1, Rule set 1	READ: always, UPDATE: never Rule for readable data that cannot be modified during personalization. A typical example of such data is the code that identifies the microcontroller type and associated memory sizes
SE1, Rule set 2	READ: always, UPDATE: ADM1, CREATE: ADM1, DELETE: never Combined rule for file access and file management. The file access rule applies to data that must be read and modified during personalization, such as name, address, date of birth and the like. Read access is necessary for verifying correct personalization in a subsequent step. The file management rule only allows data entry, since data deletion is not necessary during personalization
SE1, Rule set 3	READ: never, UPDATE: ADM1 Rule for non-readable data that can be written during personalization. An example of such data is a seed value for a random number generator or a key for encrypting keys stored in the card
SE2, Rule set 1	READ: always, UPDATE: never, CREATE: never, DELETE: never Combined rule for file access and file management. The file access rule applies to data that can be read freely but can never be modified after being stored. An example is the card number. The file management rule excludes creating new files and deleting files
SE2, Rule set 2	READ: PIN, UPDATE: PIN, CREATE: ADM2, DELETE: ADM2 Combined rule for file access and file management. The file access rule applies to user data that can be read and modified after successful PIN verification. The file management rule permits creating and deleting files after successful verification of the PIN (ADM2) of the administrative entity
SE2, Rule set 3	READ: PIN, UPDATE: ADM2, CREATE: ADM2, DELETE: ADM2 Combined rule for file access and file management. The file access rule applies to data that can only be read by the user and can only be modified by the system operator. An example of such data is the access privileges in a card used for computer access. The file management rule permits creating and deleting files after successful verification of the PIN (ADM2) of the administrative entity
SE2, Rule set 4	READ: never, UPDATE: never This rule applies to data that is only used internally by the smart card operating system and cannot be read or written by the outside world
SE2, Rule set 5	READ: PIN, UPDATE: never This rule applies to data that can be read after successful verification but can only be written internally by the smart card operating system

Besides the access rules for the data files, a variety of other conditions must be defined for each application and entered in the EF_{ARR} file. They are the conditions for creating (CREATE), deleting (DELETE), resizing (RESIZE), blocking (INVALIDATE), unblocking (REHABILITATE), and permanently blocking (LOCK) data files (EFs) and directories (DFs).

For the sake of completeness, it should be mentioned here that the rules in an EF_{ARR} file can also be used to manage a significantly larger range of privileges. For instance, it is possible to specify in such rules that Secure Messaging must be used for reading or writing certain files. This even makes it possible to specify whether data must be transmitted in authenticated or encrypted form.

For security reasons, the access conditions should always be specified as conservatively as possible. However, you must take care to ensure that suitable tests can still be performed after completion of the manufacturing phase in order to ensure correct personalization. These tests are usually based on reading or using personalization data (for example, for an authentication).[1]

Similar considerations apply to accesses that are necessary for analysing complaints about cards in the field. The access privileges should at least be sufficiently lenient to make analysis of the problem possible, but they should not create any opportunities for attacks.

5.3.2   File names

There are few restrictions on the file names (FIDs – file identifiers) for data files. The reserved FIDs specified in ISO/IEC 7816-4 are '3F00' for the root directory (MF) (master file), '3FFF' for selecting a file using a path name, and 'FFFF' for future use.

From practical experience, it is a good idea to use the same upper byte for all FIDs assigned to a set of related files. The lower byte can then take the form of an incrementing number. For example, you could assign FIDs in the range 'A001'–'A004' to the files of an access control application and FIDs in the range 'B001'–'B008' to the files of a payment application in the same smart card.

If a new application is generated for a multiapplication card, it goes without saying that the FIDs of the new application must be different from the FIDs of the applications already present in the card and all FIDs reserved for future use.

5.4   Log Files

Various forms of data that must be archived for a certain length of time are normally generated during transactions between smart cards and the outside world. Such data is typically stored in log files in smart cards. The intended use of the logged data may be to provide information to the card user or the application operator. The intended use has a significant impact on the content of the logged data. Log files are very often used in electronic purse cards to record loading and payment transactions initiated by the card user. The essential information about the logged transactions can then be displayed using a portable card terminal.

5.4.1   Data storage

EFs with cyclic structure are the best choice for storing log data. Such files have a certain number of equal-length records and one log data set is stored in each record.

---

[1] See Section 7.1

Each cyclic EF also has a pointer managed by the smart card operating system that marks the most recently written record. This mechanism can be used to implement log files such that the oldest record is automatically overwritten by the newest record, all without any additional administrative overhead. However, it should be ensured that the previous log record is completely erased (and thus set to a defined state) before the new record is written.

As a fundamental principle, only essential data should be logged in order to minimize the use of memory space and ensure rapid execution of the associated commands.

5.4.2   Assigning data to log files

Within an application, it is often necessary to log a wide variety of data items generated at different rates. It is thus useful to always store related data items in a file provided for that purpose, as otherwise data logged for infrequent types of transactions will be overwritten regularly by data logged for frequent types of transactions.

For instance, it is common practice in electronic purse applications to use separate log files for load transactions and payment transactions. Naturally, a single log file could be used for both types of transactions, with the records labelled accordingly. However, an enormous log file would be necessary to ensure that log data for a certain number of load transactions was always present in the file, since payment transactions are much more frequent than load transactions. For this reason, the log data is stored in different files according to its origin.

5.4.3   Invoking logging

Log data can be generated when a single command is processed or when a set of associated commands is executed. The latter situation is called a *transaction*.

In the case of a single command, the entire record is usually written to the log file in a single operation. As a rare exception, the record may be written in several steps during the processing of a relatively complex command. This is similar to what happens during a transaction, in which case data associated with each command is added to the record incrementally when the command is executed.

Besides automatic recording of log data during processing of commands or transactions, logging can also be initiated by the terminal. In this case, the data to be logged is transferred from the terminal to the smart card and then stored in a log file. To prevent unauthorized overwriting of the log file, the terminal should be authenticated with respect to the smart card before this is done. This mechanism is often used in offline systems to enable the terminal to transfer information to the background system indirectly via an online terminal used at a later point in time.

5.4.4   Access conditions for log files

In the simplest case, the required security condition for reading log files can be set to 'Always', which means the files can always be read with the usual commands (typically

READ RECORD). This makes it possible to use simple (and thus inexpensive) card terminals to read the contents of log files and show them on a display. As it is not necessary to enter a PIN in this case, only one button is needed on a simple pocket card reader to select and display successive records. Omitting any restrictions on read access to the log files is thus an ideal solution in terms of ease of use.

However, with regard to protection of personal data[1] it is preferable to not permit unrestricted read access to log files. Otherwise, data that certainly can be regarded as sensitive data can be ferreted out using any terminal. For example, a merchant could conceivably read log data when an electronic purse card is used to make a payment in order to discover when and where the customer used the card recently to make other payments.

To avoid such a situation, successful PIN verification could be specified as a prerequisite for reading log files. This would effectively prevent unauthorized reading of log data. However, this solution is not especially satisfactory from a technical perspective because successful PIN verification is already a prerequisite for using certain applications. Entering the PIN of the application would thus automatically enable read access to the log files at the same time. This could be avoided by specifying successful verification of a special PIN as the condition for reading log files, but this would require a second PIN for the application in question. This approach is not acceptable in terms of user acceptance. It is quite common practice to allow log files to be read at all times, even though this makes it relatively easy to ferret out personal data from smart cards.

The access conditions for log files should be formulated such that the data is fully protected against external manipulation. This ensures that the logged data is authentic. For this reason, the access condition for UPDATE should always be set to 'Never'. This ensures that the log data can only be written by the actual application, but not from outside.

5.4.5 Logged data

Table 5.5 shows a possible structure for a log file. The purpose of a log file is to allow all essential aspects of a transaction to be reconstructed at some later point in time. One of the important pieces of information for this purpose is the terminal used for the transaction. The unique identification code of the terminal (normally called the *terminal ID*) is usually sufficient for this purpose. A side effect is that the location of the transaction can also be determined if the terminal is not a mobile terminal.

Another important piece of information is a number that uniquely identifies the transaction in question, which is usually called the *transaction number*. Recording the time of day and date in each record is also helpful for manual analysis of the log data. However, smart cards do not support real-time clocks, so only the date and time provided from the outside world (usually by the terminal) can be used for this purpose. This information is not truly authentic and is not necessarily concurrent with the transaction because it may reach the smart card after a delay.

Another rule is that the essential transaction data only should be stored. In the case of an electronic purse, this would be the current balance of the purse, the amount deducted

---

[1] See Section 4.1

Table 5.5   A typical log file structure that can be used as a template for a variety of smart card applications

File:	EF$_{\text{Logfile}}$	
FID: SFI:	'2F2A' '05'	
Structure: File size:	Cyclic 30 bytes 10 records = 300 bytes	
Access conditions:		
	READ: UPDATE:	Always Never
Position	Length	Data Element
1	1 byte	Status of the log record
2	1 byte	Transaction type
3	2 bytes	Transaction number
5	4 bytes	Terminal ID
9	x bytes	Essential data from communications for transaction step n
$9 + x$	y bytes	Essential internal intermediate results of the smart card for transaction step n
$9 + x + y$	4 bytes	Transaction date from terminal (format DDMMYYYY)
$13 + x + y$	3 bytes	Transaction time from terminal (format HHMMSS)
$16 + x + y$	2 or 4 bytes	Optional error detection code (EDC) or signature (CCS)

from the purse or loaded into the purse, and the signatures associated with the transaction. The data can be regarded as complete if the essential functional and security portions of the transaction can be reconstructed from the log data and the data stored in the smart card used for the transaction.

5.4.6   Consistency and authenticity of log data

Several other data elements are used to manage records in log files. In the case of transactions involving multiple commands, the full log record is not written in a single operation, but instead a portion of the record is written for each command. A data element that indicates the current status of the transaction is thus necessary. The status indicates the actual state of the log record, from creation of a new log record to successful completion of the processing of the individual commands. Some typical states in the order of occurrence are 'New log record created', 'Step n initiated', 'Step n completed' (which means that the intermediate results for step n have been written to the log record), and 'Log record completed'.

If it is necessary for log records to always be consistent and free from truncated data, even in the event of an unforeseen power interruption, the individual data elements and associated status information must always be written using atomic processes.[1] However, this increases the execution time, so many applications do not use atomic write operations for log files.

---

[1] See Section 2.4.2

Nevertheless, there should be some easy way to determine whether a log record has been truncated. One way to do this is to calculate and store an error detection code for the entire record. This makes it possible to check each record for consistency quickly and with relatively little effort.

If the log data is not written using atomic processes, which is quite often the case, the data may not be stored completely if an unforeseen power interruption occurs. Such an interruption can be instigated by yanking the card out of the terminal during the transaction. This means that log data that is not written using atomic processes must be checked for consistency before it is displayed at the user interface, in order to avoid presenting truncated data to the user.

Depending on the specific application, it can also be worthwhile to compute a signature for each log record.[1] This makes it possible to check the authenticity of the data in the log file later on, since any manipulation can be recognized using the signature. A symmetric cryptographic algorithm, such as Triple DES or AES, is typically used to generate the signatures. This reduces the signature length, but it has the disadvantage that a secret key is necessary for checking the signatures. The signatures should be card-specific so that the records can be uniquely associated with individual smart cards.

5.4.7 Log file size

The amount of memory available to applications in smart cards is highly restricted. Log files should thus occupy only as much memory as is absolutely necessary. Transaction-based applications, such as electronic purses, typically have log files with a capacity of 10 records. The number of entries should also take the transaction type into account, since load transactions (for instance) are significantly less common than payment transactions. Similar types of transactions are often combined for the sake of simplicity. For instance, a 'Load' log file can be used for loading, unloading and cancellation transactions, while a 'Payment' log file can be used for payment, reversal and cancellation transactions.

It is quite common for a different number of log records to be recorded for different types of transactions. In the case of the Geldkarte electronic purse system used in Germany, 15 log records are stored for payment transactions and 3 records for load transactions. Maintaining records for a large number of transactions does not necessarily yield increased clarity for users, since their memories also reach back only a certain length of time.

In the case of smart cards used in partially offline systems, the application operator may have other needs and expectations regarding log files. Application operators may need additional information about transactions performed offline and may wish to read this information during online transactions. In such cases, it is important to know the typical number of offline transactions or even the maximum possible number of offline transactions, and the number of log records must be chosen accordingly.

---

[1] The term *signature* is not entirely correct in this context, since a symmetric algorithm is used. Properly speaking, it should be called a *cryptographic checksum* (*CCS*). However, the term *signature* has been used for many years in major smart card standards (such as EN 1456) as a common name for this sort of check value, so it is thus used here as a concession to general usage

5.4.8 Logging process

Sequence Chart 5.1 shows the general features of the communication process for the transactions and related activities for creating and writing log records. After the first command is received, a new record with its contents erased is created in the log file, which has a cyclic structure. The status is then set to 'Transaction step 1 initiated' and the data received from the terminal for this step of the transaction is written to the record along with the intermediate internal results from the smart card.

IFD (Terminal)		ICC (Smart Card)
Command 1	$\longrightarrow$	. . .
		Create record and erase contents
		Write status 'Step 1 initiated'
		Write log data for step 1
		Write status 'Step 1 completed'
		. . .
	$\longleftarrow$	Response 1
Command 2	$\longrightarrow$	. . .
		Write status 'Step 2 initiated'
		Write log data for step 2
		Write status 'Step 2 completed'
		. . .
	$\longleftarrow$	Response 2
	. . .	
Command n	$\longrightarrow$	. . .
		Write status 'Step n initiated'
		Write log data for step n
		Calculate and write the EDC or
		signature of the data (without the
		status)
		Write status 'Step n completed'
		. . .
	$\longleftarrow$	Response n

Sequence Chart 5.1   General form of the communication process between a terminal and a smart card and the associated operations for writing log data in a transaction with n steps

The write operation can be performed in the conventional manner or as an atomic process. The remaining parts of the command are then processed, the corresponding status is set in the log record (i.e. 'Transaction step 1 completed'), and the response is sent to the terminal.

This process is repeated until the final command has been fully processed. The status is then set to the value that marks completion of the transaction, and depending on the structure of the log file a checksum or signature of the log data is added to the record.

5.5 Pairing

In certain applications, it is necessary to forge a bond between a terminal and a smart card. Once this has been done, it is no longer possible to interchange the terminal or smart card. Many different terms are used for this process, including 'marriage', 'pairing', and 'imprinting'. The latter term is taken from the behavioural phenomenon first described by Konrad Lorenz.[1]

One method that can be used to pair a terminal with a smart card is described below. Other methods are also possible, such as the SIM lock method used for GSM cards, but the scheme described here can be implemented in either mutual or unilateral forms. The relationship between terminal and a smart card has two aspects: the actual pairing process, which is performed only once, and checking for pairing based on this process, which can be performed whenever necessary.

In the pairing process, the terminal reads a unique, non-modifiable identification number (ID_{IFD}) from memory. The terminal also generates a random number (RND_{IFD}) and stores it in its memory in unmodifiable form. The identification number and the random number are then sent to the smart card, which also reads its own identification number (ID_{ICC}) and generates a random number (RND_{ICC}). These four data elements are then concatenated and the hash value of the data is calculated. This hash value is stored in the smart card. The smart card then sends its identification number and random number to the terminal, which also calculates and stores the hash value of the concatenated data. This completes the pairing process, which is shown in detail in Sequence Chart 5.2.

IFD (Terminal)		ICC (Smart Card)
Read ID_{IFD} from memory		
Generate and store RND_{IFD}		
Command [ID_{IFD} ∥ RND_{IFD}]	$\longrightarrow$	Read ID_{ICC} from memory
		Generate and store RND_{ICC}
		$X_{ICC} = H(ID_{IFD}$ ∥ RND_{IFD} ∥ ID_{ICC} ∥ $RND_{ICC})$
		Store X_{ICC}
$X_{IFD} = H(ID_{IFD}$ ∥ RND_{IFD} ∥ ID_{ICC} ∥ $RND_{ICC})$	$\longleftarrow$	Response [ID_{ICC} ∥ RND_{ICC}]
Store X_{IFD}		

Sequence Chart 5.2   A process for pairing a terminal with a smart card. ID_{IFD} and ID_{ICC} are the identification numbers of the terminal and the smart card, while RND_{IFD} and RND_{ICC} are the corresponding random numbers generated and used in the process

The process shown in Sequence Chart 5.3 can be used to check whether a terminal and a smart card are bonded by pairing. The terminal first reads its identification number ID_{ICC} and random number RND_{ICC} from memory and sends them to the smart card. The smart card in turn retrieves its identification number ID_{ICC} and random number RND_{ICC}

---

[1] In 1930, Konrad Lorenz discovered that goslings regarded the first living being they saw after emerging from the egg as their mother for the rest of their life

IFD (Terminal)		ICC (Smart Card)
Read ID_{IFD} and RND_{IFD} from memory		
Command $[ID_{IFD} \parallel RND_{IFD}]$	$\longrightarrow$	Read ID_{ICC}, RND_{ICC} and X_{ICC} from memory
		$X_{IFD} = H(ID_{IFD} \parallel RND_{IFD} \parallel ID_{ICC} \parallel RND_{ICC})$
		IF $(X_{ICC} = X_{IFD})$ THEN
		'The terminal is the right partner'
		ELSE
		'The terminal is not the right partner'
Read X_{IFD} from memory	$\longleftarrow$	Response $[ID_{ICC} \parallel RND_{ICC}]$
$X_{ICC} = H(ID_{IFD} \parallel RND_{IFD} \parallel ID_{ICC} \parallel RND_{ICC})$		
IF $(X_{IFD} = X_{ICC})$ THEN		
'The smart card is the right partner'		
ELSE		
'The smart card is not the right partner'		

Sequence Chart 5.3   Process for checking for pairing of a terminal and a smart card

from memory. The smart card then calculates the hash value of these data elements and compares it with the originally calculated hash value. If the two values match, the terminal is the terminal from the pairing process. After this, the terminal carries out steps similar to those in the smart card to check whether the smart card is the 'right one'.

In a highly generalized sense, each of the identification numbers used in this scheme is a reference to the particular terminal or smart card involved in the paired relationship, while each of the random numbers is a reference to the specific session in which the pairing was performed.

The advantage of this method is that it is not possible to deduce a particular terminal or smart card from a hash value, since it is not possible to calculate the original data from a hash value. This means that the system cannot be broken if (for example) the hash value can be read from the terminal. This method can also be implemented in a unilateral form by omitting the steps for checking the smart card or the terminal. The unilateral form can be used to implement systems in which certain smart cards can only be used with a single, specific terminal and do not work with any other terminals.

5.6   Protecting Transaction Data

The usual way to protect transaction data transferred between a terminal and a smart card is to use the Secure Messaging mechanisms.[1] These methods are specified in the

---
[1] See Section 2.3.4

ISO/IEC 7816-4 standard and are supported by many smart card operating systems. However, sometimes it is necessary to protect the data at the application level instead of at the transmission level. This can be done using cryptographic methods similar to those used at the transmission level.

Sequence Chart 5.4 shows some commonly used methods for application-level protection of transactions between a terminal and a smart card. These methods can be combined to suit specific requirements. However, you should bear in mind that transaction protection makes systems more complex and costly. You should thus aim for a solution that adequately fulfils the requirements, rather than aiming for a solution that incorporates the maximum possible number of protective mechanisms without good reason.

All protective mechanisms involve appending various types of administrative data to the actual user data. In the interest of compatibility with future modifications, it is highly advisable to always identify all data in a manner that allows structural changes to be made with little or no impact on the program routines that process the data. The commonly preferred solution is TLV structuring of all data concerned.

In the simplest case, which is shown as Variant 1 in Sequence Chart 5.4, the command data is sent to the smart card as plain text without any supplementary cryptographic protection and the response is also sent back without protection. In this case, the command and associated response can be manipulated at will by inserting, deleting or modifying the transferred data. This option is the easiest to implement because it does not require any sort of cryptographic algorithm or associated key management and it does not require any additional processing time for cryptographic protection.

If it must be possible to detect manipulation of the transferred data, this can be achieved by appending an MAC (message authentication code) to the data. A symmetric cryptographic algorithm (such as AES) and an associated secret key are required to compute the MAC. The recipient of a message protected by an MAC can be sure that the sender is authentic because the sender possesses the secret key used to compute the MAC. This is shown in Variant 2 of Sequence Chart 5.4.

However, Variant 2 does not provide any protection against replaying previous messages and their associated MACs. This can be prevented by individualizing the messages. This can be done in two different ways. The first way is to prefix the message sent to the smart card with a sufficiently long random number (typically 4 bytes), which is also used to generate the response from the smart card. The MAC is computed over the combination of the data and the random number, so any attempt to replay previously sent messages can be detected immediately and blocked. The alternative is to use a send sequence counter (SSC) to individualize messages with a number that is incremented by 1 for each message. In contrast to individualization using random numbers, a send sequence counter also provides protection against deletion of entire messages during a communication session, since it will create a gap in the sequence of numbers. This is shown in Variant 3 of Sequence Chart 5.4. Using a send sequence counter to protect communications also provides a simple means to authenticate the smart card or the terminal with respect to the opposite party.

In systems in which the MAC generated by the smart card must be checked by a background system as well as by the terminal, both parties can use the appended MAC for this purpose in the simplest case. However, if you have to take into account the

IFD (Terminal)		ICC (Smart Card)
Variant 1		
Command [DK]	$\longrightarrow$	. . .
	$\longleftarrow$	Response [DA]
Variant 2		
$MAC_{DK} = M(DK, K_{MAC})$		
Command [DK —— MAC_{DK}]	$\longrightarrow$	. . .
		$MAC_{DA} = M(DA, K_{MAC})$
	$\longleftarrow$	Response [DA —— MAC_{DA}]
Variant 3		
SSC = SSC + 1		
$MAC_{DK} = M((SSC$ —— $DK), K_{MAC})$		
Command [SSC —— DK ——	$\longrightarrow$	. . .
MAC_{DK}]		
		$MAC_{DA} = M((SSC$ —— $DA), K_{MAC})$
	$\longleftarrow$	Response [DA —— MAC_{DA}]
Variant 4		
$MAC_{DK} = M(DK, K_{MAC})$		
Command [DK —— MAC_{DK}]	$\longrightarrow$	. . .
		$MAC1_{DA} = M(DA, K1_{MAC})$
		$MAC2_{DA} = M(DA, K2_{MAC})$
	$\longleftarrow$	Response [DA —— $MAC1_{DA}$ ——
		$MAC2_{DA}$]
Variant 5		
$EDC_{DK} = C(DK)$		
SSC = SSC + 1		
encDK =		
$E(SSC$ —— DK —— EDC_{DK}, K)		
Command [encDK]	$\longrightarrow$	. . .
		$EDC_{DA} = C(DA)$
		encDA = $E(SSC$ —— DA ——
		EDC_{DA}, K)
	$\longleftarrow$	Response [encDA]

Sequence Chart 5.4 Typical methods that can be used at the application level to ensure the authenticity of transactions between a terminal and a smart card. For the sake of simplicity, the padding required for block-oriented cryptographic algorithms and TLV coding of the transferred data (which is advisable) are not shown

possibility that a successful attack could be mounted against the secret key in the terminal, a second MAC can be used. This additional MAC is computed in the same manner as the first one, but the secret key that the smart card uses to generate the second MAC is known only to the background system (and thus not to the terminal). With this approach, it is still possible to check the authenticity of the message even if the MAC key of the terminal has been broken. This is shown in Variant 4 of Sequence Chart 5.4, in which supplementary protection mechanisms such as a send sequence counter or encryption have been omitted for the sake of simplicity.

If the transferred data must be treated as confidential, it must be transferred in encrypted form. In addition, the recipient of the encrypted data must be able to unambiguously ascertain that the received data has been decrypted correctly. This can be ensured by using a data structure that can be verified by the recipient or by using an error detection code (EDC), which is calculated from the data and appended to it as shown in Variant 5 of Sequence Chart 5.4. In principle, all types of hash algorithms and error detection codes are suitable for this purpose, such as a CRC (cyclic redundancy code) or a Reed–Solomon EDC.

5.7   Reset-proof Counters

A reset-proof counter always counts in a particular direction, even if its operation is interrupted by a reset or an abrupt smart card power-down. Counters of this type are used when it is essential to ensure the reliability of the counting process. A typical example of a reset-proof counter is the counter for the number of authentications of a smart card. In this case, it is important to ensure that the specified maximum number of authentications is never exceeded under any circumstances.

The easiest way to create a reset-proof counter is to implement it in EEPROM and protect the counting process by implementing it as an atomic transaction. The disadvantage of this method is that it imposes a heavy load on the EEPROM and the counting process takes considerably longer than it would with a counter implemented in RAM, because of the relatively long erase and write times of the EEPROM. However, a counter that operates in volatile RAM memory can be repeatedly reset to its initial value by a smart card reset or a power-down/power-up process. This would make such a counter useless.

This problem can be solved by periodically backing up the count in nonvolatile memory, such as EEPROM. Figure 5.1 shows this process in detail. In the case of an up counter, the number of time-intensive backup operations can be reduced by rounding up the actual count before it is backed up and storing the rounded-up value in nonvolatile memory. This makes it impossible to reset the counter to its initial value by generating a reset during the counting process. The counter can only assume the rounded-up value after a reset, which is always disadvantageous for an attacker. The stored count is only updated when the rounded-up value of the actual count exceeds the stored value, and the backup operation should be protected against interruptions by implementing it as an atomic transaction. In the case of a down counter, this process must be adapted accordingly.

The count is re-initialized after a reset by copying the stored count from EEPROM into RAM. This means that the counter cannot be forced to assume a previous value by external manipulation involving a reset or power-down process. All forms of external manipulation will cause the counter to attain its final state prematurely, which is by no means in the interest of an attacker.

5.8   Proactivity

The transmission protocols used with smart cards comply with a strict master–slave principle, with the terminal acting as the master and the smart card as the slave. As

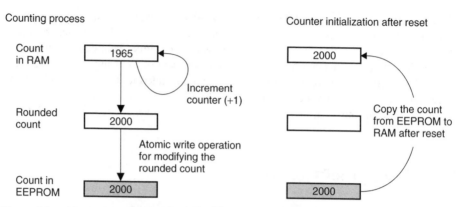

Figure 5.1 Sample implementation of a reset-proof counter. The counting process is shown at the left, while the process for initializing the counter after a reset is shown at the right. In this example, the count is always rounded up to an integer multiple of 100 units

a consequence, all communication can only be initiated by the terminal. There is no possibility for the smart card to assume an active role.

However, there are applications in which it must be possible for the smart card to initiate a communication sequence. This basically involves all applications in which the smart card needs to assume control of a terminal or some other type of equipment.

To ensure continued compatibility with existing transmission protocols, there is a method that allows commands originating from the smart card to be superimposed on the normal master–slave protocol. This is called *proactivity*. Proactivity in the smart card world was developed and introduced in the 1990s as part of the SIM Application Toolkit (SAT) for SIM cards used in the GSM mobile telecommunication system.[1] The following paragraphs and Sequence Chart 5.5 describe the general command processes used for proactivity, with close reference to the SIM Application Toolkit used in the GSM system.

The underlying mechanism of proactivity is periodic querying of the smart card by the terminal to determine whether the smart card wishes to send a command to the terminal. In terms of information technology, this is a form of polling. This querying should be simple and take up only a small amount of transmission time. The Case 2 command STATUS is typically used for this purpose.

If the response to a STATUS command indicates that the smart card wishes to send a proactive command to the terminal, the terminal uses the FETCH command to retrieve the command. The proactive command is transferred in the data portion of the response to the FETCH command. Now the proactive command can be processed in the terminal. When the response to the command is ready, it is sent to the smart card using the TERMINAL RESPONSE command. This completes the command–response cycle for a proactive command. The terminal then continues to issue periodic queries using the STATUS command.

A few aspects of this generic approach to proactivity deserve special attention. The proactive commands transferred in the data portion and their associated responses should

---
[1] Described in detail in TS 51.014 (2003)

IFD (Terminal)		ICC (Smart Card)
Cyclic query		
STATUS	$\longrightarrow$	. . .
		X = proactive command available or not available
	$\longleftarrow$	Response [X]
Sequence for proactive command		
FETCH	$\longrightarrow$	. . .
	$\longleftarrow$	Response [Command$_{IFD}$]
Process the proactive command		
TERMINAL RESPONSE	$\longrightarrow$	. . .
[Response$_{IFD}$]		
	$\longleftarrow$	Response

Sequence Chart 5.5   The first part of this sequence chart shows cyclic querying of the smart card by the terminal to determine whether the smart card has a proactive command ready to send to the terminal. The second part portrays the process of retrieving a proactive command, processing it in the terminal, and returning a response to the smart card. The process shown here assumes a positive result

always be embedded in a TLV data structure in order to facilitate changes and extensions. It is also important to specify the cycle time for the periodic queries at the very beginning of the proactive session. This can best be done by the first proactive command sent to the terminal, which thus enables the smart card to independently determine the value of this important time parameter. The cycle time essentially determines the response time of the combination of the smart card and the terminal, and it must be specified individually for each application. It is typically set to a value of 20 s in the SIM Application Toolkit.

Proactivity makes it possible to reverse the rigid one-directional relationship of the master–slave principle employed in smart card transmission protocols without sacrificing compatibility with existing processes. This technique has been used successfully for many years in the SIM Application Toolkit. Proactivity can also be expanded to include other useful functions. Some examples are proactive commands for setting timers or defining specific events in the terminal that will trigger sending a message to the smart card. This makes it possible to circumvent the rigid, periodic timing imposed by cyclic transmission of the STATUS command.

5.9   Authentication Counter

Unilateral challenge–response authentication is normally used to authenticate smart cards in a system. For this purpose, the background system sends a random number to the smart card, which encrypts the random number and returns the result. The same operation is then performed in the background system and the result is compared with the result received from the smart card. If the results of the computations performed by the smart card and the background system are the same, the smart card knows the shared secret key and is thus authentic from the perspective of the background system. This process is used to authenticate SIM cards in the GSM mobile telecommunication system, for example.

The number of authentications is usually not limited, so the authentication process can be used to have a smart card encrypt a large number of randomly chosen numbers. This provides a good basis for attacks on the cryptographic algorithm. Such attacks could involve analysing the current consumption of the smart card during encryption or analysing the plain-text–cipher-text pairs obtained from the authentications. The latter method is used in attacks on SIMs that employ the COMP 128-1 cryptographic algorithm. The key used with COMP 128-1 is too short, so the algorithm can be broken, which means the secret key can be determined by computation if the attacker has approximately 60 000 plain-text–cipher-text pairs to work with.

Changing the cryptographic algorithm used in a large smart card system is difficult, and in many cases it is too costly for the system operator. For these reasons, the relatively weak COMP 128-1 cryptographic algorithm was not systematically replaced in GSM systems after this successful form of attack became known, but instead is still in widespread use in many countries. However, several measures were developed to defend against the attack described above.

IFD (Terminal)		ICC (Smart Card)
Command [RND]	$\longrightarrow$	Fetch Z and Z_{max}
		IF $(Z \geq Z_{max})$ THEN abort
		$Z = Z + 1$
		Store Z
		$X = E(RND, K)$
	$\longleftarrow$	Response [X]

Sequence Chart 5.6   An example of using a counter for challenge–response authentication of a smart card. The random number RND is passed in the challenge to the smart card, which first checks whether the maximum number of authentications has been reached. If this is the case, further command processing is terminated and a suitable response is sent to the terminal. If counter Z has not yet reached its maximum value Z_{max}, the random number RND is encrypted using key K and the response X is returned to the terminal in response

To prevent attacks based on current analysis and logical analysis, measures were implemented to stop attackers from performing a large number of authentications, since both forms of attack require a large number of plain-text–cipher-text pairs. For this purpose, the command used for authentication (typically INTERNAL AUTHENTICATE) was extended to include a use counter that is incremented by 1 for each authentication. If this counter reaches a limit value stored in the card, all further authentications are blocked or essential functions of the smart card are disabled. When this happens, the terminal may (for example) receive a special return code indicating that the maximum number of authentications has been reached. It is important to specify the maximum value such that the number of authentications is not sufficient for analysis but is still large enough to enable the smart card to be authenticated for the duration of its anticipated service life.

If the authentication counter is implemented using a file that permits write access for an entity having administrator privileges, it can be reset if necessary after it reaches the maximum value. This can be useful in cases where it is reliably known that the maximum number of authentications was reached in normal use rather than due to some form of attack. Another possibility is to switch automatically to a different key after the maximum value is reached, which makes the already ferreted out plain-text–cipher-text pairs worthless.

It would also be possible to recognize an attack based on a large number of authentications by analysing the pattern of the challenges, since they are simply incremented in most such attacks. However, on the basis of security considerations it is always advisable to restrict the number of authentications to a reasonable value.

5.10   Manual Authentication of a Terminal

In normal life, smart cards are often used with terminals that are unfamiliar to the user. In such cases, it is usually not apparent to the card user whether the terminal is genuine or fake. This is especially critical with regard to PIN entry,[1] since fake terminals are often used to spy out PIN codes. However, there are simple methods that can be used to determine whether a terminal is genuine.

This requires making a few modifications to the application in the smart card. Essentially what is required is a new file that can be accessed for reading only after the smart card has successfully authenticated the terminal. The content of this file can only be modified if the terminal has been authenticated and the user been identified by a successful PIN verification. A challenge–response process is normally used to authenticate the terminal. Table 5.6 shows the content of the file used by the terminal authentication process and the necessary access conditions.

Table 5.6   Structure and access conditions of the EF$_{US}$ file used for manual authentication of a terminal

File:	EF$_{UserSecret}$	
Structure:	Transparent	
File size:	x bytes	
Access conditions:		
	READ:	Terminal authentication
	UPDATE:	(Terminal authentication) AND (PIN verification)

Position	Length	Data Element
1	x bytes	User secret

The life cycle of this application extension consists of the usual two phases. During the administrative phase, a secret word known only to the card user is stored in the User Secret EF (EF$_{US}$) in a one-time operation. This can be done using a special administration

---

[1] See Section Section 9.7

terminal that the user trusts, equipped with a normal computer keyboard. Write access to this file is only granted after successful authentication of the terminal and successful verification of the user's PIN code. The administrative phase ends when the user secret is stored in the smart card. After this, the smart card can be used with any desired terminal to check whether the terminal is genuine.

The process for manual authentication of a terminal is described below and shown schematically in Sequence Chart 5.7. Immediately after the smart card has been inserted in an unfamiliar terminal and communication has been established, the terminal issues the GET CHALLENGE and EXTERNAL AUTHENTICATE commands to attempt to authenticate itself relative to the smart card. If this is completed successfully, the smart card allows read access to the file containing the user secret. The terminal then uses READ BINARY to read the content of this file and show it on the display. The user sees the secret word, which only he or she knows, and can thus be certain that the terminal is authentic. The user does not enter his PIN, which is critical for secure use of the smart card, until after this has happened.

IFD (Terminal)		ICC (Smart Card)
GET CHALLENGE	$\longrightarrow$	Generate random number
X_{IFD} = enc(Random number, Key)	$\longleftarrow$	Response [Random number]
EXTERNAL AUTHENTICATE [X_{IFD}]	$\longrightarrow$	X_{ICC} = enc(Random number, Key)
		IF (X_{IFD} = X_{ICC})
		THEN Status = T_{auth}
		(Terminal authenticated)
		ELSE Status = T_{nauth}
		(Terminal not authenticated)
IF (Status = T_{nauth}) THEN abort	$\longleftarrow$	Response [Status]
READ BINARY [EF_{US}]	$\longrightarrow$	IF (Status = T_{auth})
		THEN read access to EF_{US} allowed,
		Y = Content EF_{US}
		ELSE read access to EF_{US} prohibited,
		Y = abort
Display [Content EF_{US}]	$\longleftarrow$	Response [Y]
Display [Enter PIN]		
User entry [PIN]		
VERIFY [PIN]	$\longrightarrow$	. . .
	$\longleftarrow$	Response [Result of PIN verification]

Sequence Chart 5.7 Typical communication sequence between a terminal and a smart card for manual authentication of a terminal. The user secret is stored in the user secret EF (EF_{US})

In principle, it would also be possible to store a graphic object or photo in the user secret EF instead of a secret word and then show this object or photo to the user after successful authentication of the terminal. However, this would require terminals with graphic displays and most terminals do not meet this requirement. Manual authentication of a terminal is an especially beneficial additional function for applications involving PIN entry at terminals that are not familiar to users.

5.11   PIN Management

Numeric codes have been used for many years to authenticate card users. Only a simple ten-digit numeric keypad is needed to enter the codes and numbers are also suitable in terms of the ability of the general population to remember them.

However, this subject requires attention to more than just the technical aspects. You also have to take the behaviour and preferences of the users into consideration. Smart cards are used in all reaches of society, so only well established and widely accepted methods, such as PIN entry, should be employed.

This is also the reason for the widely used PIN code length of four digits. Although the theoretical security of PIN codes increases with the number of digits, the practical security reaches a maximum at four digits. If a larger number of digits is used, more users will either write the PIN code on the card or keep it in a handy location near the card. The number of cards that are blocked because of incorrect PIN entries also increases in proportion to the length of the PIN code, with a corresponding decline in user satisfaction and significantly increased administrative costs. Even now, complaints due to forgotten PIN codes are by far the most common cause of customer service events.

The established PIN code lengths – four digits for normal functions and six or at most eight digits for special functions – have been proven in practice and should be used in new smart card systems. PIN codes for special functions, such as changing a blocked PIN code, can be longer because they are usually read from a PIN letter located in a file somewhere in the user's home.

The PIN error counter normally blocks the application in the smart card after three incorrect PIN entries. This is also regarded as tolerable with regard to security. In the case of longer PIN codes for special functions, the maximum value of the error counter before blocking occurs can be increased to as much as 10 for some applications. In this regard, it is necessary to pay attention to a trivial but nevertheless often overlooked aspect: there must be a function present to allow the error counter to be reset to its initial value after it has reached its maximum value. If this function is overlooked, the smart card will be useless after the maximum error count has been reached.

The reset function can be implemented individually in each card by using a personal unblocking number (PUK). This requires the user of the smart card to enter the PUK and his new PIN in the smart card in a single session. A new PIN is necessary because the user has obviously forgotten his previous PIN. In principle, it would also be possible to reset the error counter after successful mutual authentication of the smart card with a terminal or the background system. This scenario does not require a PUK, but the user must still specify a new PIN code. There are also applications that use a down counter to limit the number of times the PIN code can be changed, although this does not provide much benefit from a security perspective.

System operators are sometimes interested in knowing whether the users of their smart cards also use terminals that do not belong to the system, in particular home terminals that can be used to figure out which functions the smart card supports. An interesting approach to this was implemented quite a few years ago in a petrol card system used in Great Britain. The official terminals allowed only three incorrect PIN entries, but the error counter in the smart card did not block the card until four incorrect attempts had occurred. In case of a complaint due to a blocked PIN code, it was thus possible to use an administration terminal to see directly whether the card user had been experimenting with the card on his own. Whether anything useful could be done with this information is another question.

Of course, this approach could be quite useful for other smart card applications under certain conditions. Nevertheless, in currently used systems the state of the error counter cannot be regarded as a reliable indication of private experimenting with the smart card, since it does not provide sufficiently strong evidence. A potential alternative would be to make an entry in a log file each time the error counter reaches its maximum value while the card is being used with a genuine terminal. This would make it possible to demonstrate quite clearly whether incorrect PIN codes were entered too often using a terminal not belonging to the system.

Modern smart card operating systems can easily manage separate PIN codes and associated PUKs for individual applications and even individual functions of an application. In theory, this gives application developers a broad range of design options. However, our advice here is that it is better to ignore these options, since otherwise ease of use ('user friendliness') is bound to be left far behind. You should adhere to the model of one PIN per card as much as possible. In certain cases, such as smart cards with an additional signature function, it is unfortunately not possible to avoid using a second PIN code. In such cases, the system operator must anticipate higher service costs due to forgotten PIN codes.

The best form of training for remembering a PIN code consistently is to use it frequently. This consideration should thus be borne in mind during application development, since the incidence of problems with forgotten PIN codes increases when PIN codes must be entered only rarely.

One option for PIN codes that unquestionably warrants consideration is free selection by users. This has indisputable mnemonic advantages, but it carries the risk that many users are inclined to choose trivial PIN codes. A simple countermeasure is to use a simple test routine in the terminal or smart card to block the usual trivial PINs, such as 0815, 1234, 4711, and four identical digits.

Allowing users to select their own PIN codes makes a system more user-friendly without significantly reducing security. However, in this case it is not possible to prevent users from choosing the same PIN code for all their smart cards and other applications, since this is not known to the system that tests the PIN codes. The seemingly obvious countermeasure of periodically changing PIN codes should be avoided as much as possible because it leads to an enormous increase in the number of incorrect PIN entries.

5.12   One-time Passwords

The simplest form of security for logging in to a computer system is entering a user name and password. However, if an attacker succeeds in spying out this information, he can log in to the user's account without authorization. This sort of attack is generally called a *replay attack*. As eavesdropping on communications is very easy with public networks in particular, various mechanisms have been developed since the early days of computer networks to defend against such attacks.

The most widely used form of defence against replay attacks at present is one-time passwords (OTPs). If the one-time passwords are numeric passwords, they are also called transaction numbers (TANs).

The procedure for using one-time passwords is as follows. Instead of entering a fixed password, the user enters his user name and a password (a number or term) that is only valid for a single transaction. This is a one-time password, and in the simplest case the user takes it from a printed list (TAN list) containing a large number of one-time passwords that must be used in sequence. The system operator sends each user a new list when all the passwords in the user's previous list have been used up. The system into which the user logs in knows the contents of the list and can compare the passwords entered by the user with the passwords in the list.

If an attacker succeeds in spying out a user name and the current password, the information will not help him because the password is valid only once. In order to benefit from the information he has spied out, the attacker must use some sort of trick such as trying to log in to the system faster than the real user, which is called a *race attack* and is generally quite difficult.

The passwords in the list must satisfy several criteria for this system to be truly secure. They must never be predictable and it must never be possible to figure out the next password even if all the previously used passwords are known. This means that they must have a nearly uniform distribution over the entire range of values, along with other characteristics of good random numbers. Of course, one-time passwords must be long enough to prevent successful access by simple guessing. The necessary length is typically on the order of five to six decimal digits in the case of numeric passwords.

In 1981, Leslie Lamport published a document[1] in which he described a possible implementation of one-time passwords. His approach consisted of two stages. In the initialization stage, a set of one-time passwords is generated using a hash algorithm and a specific initial value, and these passwords are then made available to the user. By contrast, the server receives only an initial value, which can be used to calculate each successive password. Therefore, it does not know all the one-time passwords provided to the user. This system is described in (RFC 2289, 1998) and is often used in PC networks. However, it has the disadvantage that each user must keep a large list of one-time passwords and must receive a new list each time the one-time passwords in the previous list have been used up. If a one-time password is lost while it is being conveyed to the server, the system loses synchronization, which unquestionably presents a problem in actual practice. The process of Lamport's scheme is shown in Sequence Chart 5.8.

Only transaction numbers (numeric one-time passwords) are used in modern systems based on smart cards. Although the Lamport scheme is widely used in the PC world, it has not become established in smart card systems. A mechanism that generates any desired quantity of transaction numbers in the smart card as necessary is used instead. The same computation is performed in parallel in the background system and the result is compared with the transaction number generated by the smart card. If the result of the comparison is positive, the user login is completed successfully.

---

[1] See Lamport (1981)

IFD (Terminal)		ICC (Smart Card)
Administrative phase		
Command [s]	$\longrightarrow$	$\mathrm{TAN}_1 = H(s)$
		$\mathrm{TAN}_2 = H(\mathrm{TAN}_1)$
		. . .
		Store $\mathrm{TAN}_1 \ldots \mathrm{TAN}_{n-1}$
		$\mathrm{TAN}_n = H(\mathrm{TAN}_{n-1})$
Store TAN_n	$\longleftarrow$	Response $[\mathrm{TAN}_n]$
Operational phase		
Command	$\longrightarrow$	Fetch TAN_{n-1}
Fetch TAN_n	$\longleftarrow$	Response $[\mathrm{TAN}_{n-1}]$
IF $(\mathrm{TAN}_n = H(\mathrm{TAN}_{n-1}))$		
THEN Authentication successful		
ELSE Authentication not successful		
Store TAN_{n-1}		

Sequence Chart 5.8 Typical sequence of events in the administrative phase and subsequent operational phase of a system using one-time passwords according to Lamport's scheme. Here s designates the seed (initial value) and TAN the current transaction number. Verification of a transaction number in the operational phase as illustrated here is shown using TAN_{n-1}

A technical implementation of this scheme is shown in Sequence Chart 5.9. The following data is stored in each smart card in the initial administrative phase: a seed value for the counter of the TAN generator, a unique numeric identification code, and a secret card-specific key. In operation, the counter value and the identification code are formed into a single data block that is then encrypted using a symmetric cryptographic algorithm with the card-specific key. The counter is incremented by 1 for each transaction. This method makes use of one of the characteristic of a good cryptographic algorithm, which is that changing a single input bit affects half of the output bits on average (the 'avalanche effect'). The cipher text is converted into a human-readable number and then used as a transaction number. For this purpose, several output bytes can be XORed together and then mapped into the normal decimal number space.

In the operational phase, the unique identification code is passed to the background along with the transaction number. This gives the background system a reference that it can use to read the card-specific key of the card concerned and the current counter state from a database and use them to recompute the transaction number of the smart card. For this reason, the background system must use a database to store the current counter state of each of the smart cards in the system.

However, it is still possible for a transaction number to be lost during conveyance, and the user may generate more transaction numbers than he actually uses. To make it possible to restore synchronization between the background system and the smart card in such cases, the system is designed to regard a certain number of transaction numbers following the currently valid transaction number as equally valid. If the background system determines that a particular transaction number is invalid, it computes a certain

IFD (Terminal)		ICC (Smart Card)
Administrative phase		
Command [ICCID, Z, K]	$\longrightarrow$	...
		Store ICCID
		Store Z
		Store K
	$\longleftarrow$	Response [OK]
Operational phase		
Command	$\longrightarrow$	...
		$Z = Z + 1$
		$TAN_{ICC} = E(Z \| ICCID, K)$
Fetch Z and K using ICCID as a	$\longleftarrow$	Response [ICCID | TAN_{ICC}]
reference		
$TAN_{IFD} = E(Z \| ICCID, K)$		
IF ($TAN_{IFD} = TAN_{ICC}$)		
THEN Authentication successful		
ELSE Authentication not successful		

Sequence Chart 5.9   Typical sequence of events in the administrative phase and subsequent operational phase of a system using one-time passwords. Here the TAN generator is based on a symmetric cryptographic algorithm. The ICCID (integrated chip card identifier) is a unique identification code for the smart card, Z is a counter, and K is a card-specific key. The process shown here assumes a positive result

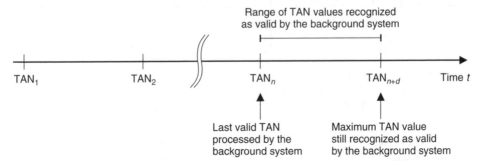

Figure 5.2   Validity of transaction numbers as a function of various counter states when the transaction numbers are generated. TAN_n is the most recent valid transaction number passed to the background system and TAN_{n+d} is the last valid transaction number that is still valid. The distance d between these transaction numbers is the range within which a transaction number will be accepted as valid

number of possible subsequent transaction numbers. If one of them is recognized as valid, the associated counter state is taken to be the new current state and is stored for future use. All counter states prior to this state and all transaction numbers corresponding to these counter states are thus rendered invalid. Figure 5.2 shows this situation plotted against a time scale.

5.13   Key Management

Key management for smart card systems encompasses an enormous variety of options. Examples that can be found in actual practice range from a single key for all system functions to highly complicated key management schemes with 30 or more derived keys for each smart card. The reasons for this wide range of variation can be found in the individual applications and the number of smart cards in the field for the specific applications.

The primary objectives of good key management are protecting the system against attackers and providing a good fallback position in the event of a successful attack. Consequently, simple smart card systems that are not especially attractive targets of attack usually have correspondingly simple key management. The most elaborate forms of key management are used in electronic purse systems and smart card systems for pay TV, both of which are unquestionably exposed to the most severe forms of attack.

Technically sophisticated key management schemes employ a different key for each function, which is called *key diversification*. The key for each function is called the *master key* for that function. On the basis of this master key, individual keys (derived keys) can be derived for each smart card and supported function. Dynamic keys and session-specific keys can in turn be generated from the derived keys. These dynamic keys and session keys are ultimately used by the cryptographic algorithms for the actual functions. Figure 5.3 shows this in graphic form.

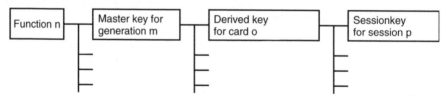

Figure 5.3   Key hierarchy of an elaborate key management system, such as is used in electronic purse card systems, with a separate key hierarchy for each smart card function. The number of key generations stored in individual smart cards is typically five or less

This means that an attacker must work his way along the entire chain, from the session-specific keys through the derived keys to the master key, to fully break the cryptographic protection of the smart card system for a particular function. To make things even more difficult for attackers, several generations of keys can be stored in each smart card so that the system can switch from one generation to the next at regular intervals or as necessary. Alternatively, means can be provided to download new keys to the smart cards from the background system. The best way to do this is to use an asymmetric cryptographic algorithm such as RSA or ECDSA.

The approach outlined above, which is described in detail in Rankl and Effing,[1] is a reasonable option for a high-security system, but it is doubtlessly excessive for normal applications in environments not overly subject to attack. Even in the GSM mobile telecommunication system, many network operators use only a single key (derived from

---

[1] See Rankl and Effing (2002)

a master key) for authentication and do not use mechanisms such as dynamization, versioning and key exchange in the cards in the field.

Key management requires key identifiers (KIDs) so that each key in the smart card can be addressed uniquely. Besides their primary role in selecting the proper keys, KIDs are necessary so that the keys can be blocked and subsequently unblocked. Unused keys present in a smart card must always be blocked to prevent attackers from having direct access to them. These management functions demand corresponding functionality, which must be provided by the smart card operating system.

On the basis of experience in the mobile telecommunication sector, a relatively simple key management scheme is a reasonable choice for applications that do not represent especially attractive targets of attack. At minimum, this must include card-specific keys for each of the functions, but more extensive measures are normally not necessary. Card numbers that are unique within the system are necessary for identifying the smart cards and with them the derived keys, but such card numbers must anyhow be available in some form for individualization of the smart cards. Although such a scheme undoubtedly increases the risk and reduces the scope of options in case of an attack, it also fosters simpler and stabler system behaviour.

The most important consideration is to comply with the unyielding principle applicable to every key management scheme, which is, that it should never be possible for someone to obtain access to the keys used in the system. This security risk must be avoided effectively, and doing so provides more protection than even the most sophisticated key management scheme with all presently known security functions.

5.14   State Machines for Command Sequences

Applications are typically implemented by specifying files and associated access privileges. These privileges can be attained by successful execution of a PIN verification or an authentication. An advantage of this approach is that it is easy to extend the functionality of the application by simply adding new files and commands to an existing application. However, it has the disadvantage that a potential attacker can systematically try out all the supported commands using any desired parameters. This can yield information that is useful for attacks, such as the allowed values of parameter bytes P1 and P2 in the command headers.

State machines for command sequences have been known since the early days of smart card operating systems. They can be used to ensure that the smart cards will only accept commands in a specified sequence with previously defined parameter values. If a command received by the smart card does not correspond to the defined sequence or contains parameter values that do not correspond to the defined parameter values, it is blocked directly by the command interpreter of the smart card and a suitable return code is sent back to the terminal. When this mechanism is used, it is no longer possible to try out commands in any desired sequence. For example, if the first operation of the sequence is specified to be authentication of the terminal by the smart card, the card will not accept any other commands until the terminal has been shown to be genuine.

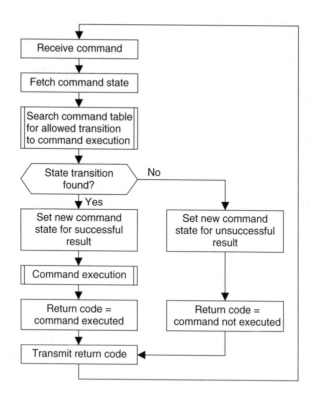

Figure 5.4   Flow chart of the main loop of a command interpreter in a smart card operating
system, showing the essential elements needed to implement a state machine for
command sequences. The variable 'command state' holds the current state of the
selected application

A state machine for commands requires a command interpreter as a central decision
authority and a table that lists the states and allowed state transitions. In addition, the
current state ('command state') is stored in volatile memory. Figure 5.4 illustrates the
basic process of the state machine in the form of a flow chart. When the smart card
receives a command, it is forwarded internally to the command interpreter and decoded
by the interpreter. The interpreter then checks to see whether an allowed transition to a
new state is available, based on the current state and the command it has just decoded.
It uses the information in the state table for this purpose. If an allowed transition is
available, the command can be executed and the command interpreter branches to the
program code for the command. If no suitable state transition is available, the command interpreter sends a suitable message back to the terminal and the command is not
processed.

Figure 5.5 shows a state diagram in which the first step is mandatory authentication
of the terminal by the smart card. This prevents attackers that are not successfully
authenticated from carrying out experiments using arbitrarily selected commands. A
possible implementation of the associated state table is shown in Table 5.7.

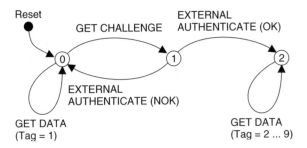

Figure 5.5   An example of a state diagram that specifies that the terminal must be authenticated before certain data objects can be read. If the authentication fails, the state machine returns to state 0. The only allowed operation in this state is to read a particular data item

Table 5.7   Example of a state table that could be used to implement the state diagram is shown in Figure 5.5. This table forms a basic reference for the command interpreter of the smart card

Command	Entry state	Next state in case of a positive result	Next state in case of a negative result
GET CHALLENGE	0	1	–
EXTERNAL AUTHENTICATE	1	2	0
GET DATA, Tag = 1	0	0	0
GET DATA, Tag = 2 . . . 9	2	2	2

The state machine principle can be extended by storing a separate command state for each application present in the smart card. This prevents the attained state of a particular application from being lost if another application is selected in between. The principle can be extended even further by maintaining a separate command state for each logical channel.[1] This makes it possible to use several applications quasi-concurrently.

State machines are commonly used to implement access conditions for files, but they can also be used to provide a supplementary firewall function, although this restricts the flexibility of the application concerned to a certain extent. Nevertheless, it is recommended to use state machines with applications that are not overly complex, work with data objects and have relatively rigid command sequences, particularly as technical implementation of state machines is comparatively easy.

5.15   Speed Optimization

Simple applications with only a few commands and small data volumes are fairly uncritical with regard to speed, since all operations are completed so quickly that the user

---
[1] See Section 2.3.5

either does not notice them or at least finds the processing time tolerable. The situation is different with applications that must handle large volumes of data or use commands that require a lot of computation. In such cases, the resulting transaction times can easily exceed the acceptable range. At this point, if not earlier, it is necessary to initiate optimization measures, although it is actually better to devote a certain amount of attention to this issue from the very start.

The essential factors that determine the execution time of a smart card application are the computation time of the processor, the data transmission time, and the read and erase times of the nonvolatile memory (EEPROM or flash). Significant speed improvements can be achieved by taking appropriate measures with respect to all three factors, without making any changes to the fundamental features of the application. As speed optimization can be a very laborious process, depending on the specific application, it is essential to generate a time analysis (profile) first. The elements that consume the most time can be identified from the profile and subsequently optimized. To achieve a satisfactory result at an acceptable cost, you should concentrate primarily on the major 'time gluttons' and avoid being distracted by minor time consumers. This is illustrated by a real-life example described below.

The example is a PC tool for processing data from USIMs, in which the first step is to read all files for which corresponding access privileges are available. The data is then displayed on the PC, where it can be modified as appropriate, after which it is written back to the smart card. As a modern USIM with a 128-KB EEPROM can easily contain more than 400 files, the read process alone takes a considerable amount of time. In its original form, the PC tool powered up the smart card and used it with the standard data transmission settings (3.5712-MHz clock rate and a divider value of 372). Each EF was selected in turn using SELECT, the file structure was determined using GET RESPONSE, and the data in the file was read using either READ BINARY or READ RECORD. The read process for all the files lasted more than three minutes, which users clearly regarded as far too long to wait.

As the first step in optimizing the read process, the communications between the terminal and the smart card were analysed. There are several commercial software tools that can perform such an analysis, although a standard protocol analyser can also be used for this purpose. The analysis showed the obvious weaknesses in the programming of the PC tool, which in turn formed the basis for subsequent improvements to the PC software.

The first optimization was to increase the clock frequency to the maximum allowed value of 5 MHz at a supply voltage of 5 V and reduce the data transmission divider to the minimum possible value of 32. This accelerated communication by more than a factor of 10 at the bit transmission level.

The next step in the optimization process was to use implicit file selection with the READ BINARY and READ RECORD commands instead of explicit file selection with the SELECT command. This was possible because the structure of each file in the USIM was defined by the TS 31 102 specification, so it was not necessary to explicitly select each file and then determine its file structure. This step made all instances of the SELECT and GET RESPONSE commands redundant.

The ultimate result of the optimization was that the read time was reduced to less than 30 s, which was an extraordinary improvement relative to the original situation. It is also worth noting that no changes at all were required to the application in the smart cards.

The following sections describe some commonly used methods for optimizing the transaction speed of smart card applications. These methods were selected on the basis of the criterion that they require only minor modifications or no modification at all to the application in the smart card.

5.15.1 Computing power

The computing power of the processor is an essential factor in the processing time of a command or a cryptographic algorithm. As the computing power is directly proportional to the clock frequency and it is relatively easy to change the clock frequency in a smart card, this option is often used as a lever to increase the computing power. For this reason, the maximum possible clock frequency should always be used. It is usually specified as 5 MHz.

For cost reasons, quite a few terminals do not have a separate clock generator for the smart card, but instead use a clock generator provided for internal use as the source of the clock for the smart card. Unfortunately, this approach often yields a clock frequency that is significantly less than 5 MHz. However, almost all modern smart card microcontrollers have an integrated clock multiplier that can be enabled if so desired and which allows the processor in the chip to be operated at a frequency as high as 60 MHz in some cases. If this option is used, you must of course ensure that the current consumption does not increase to a level that cannot be supplied by the terminal. The maximum current consumption is often strictly limited in mobile terminals, for example to 10 mA. In such cases, microcontrollers with internal hardware that automatically adjusts the internal clock frequency to avoid exceeding a specified maximum current level are a good choice.

5.15.2 Communication

The speed of the communication between the terminal and the smart card is essentially determined by the supplied clock frequency and the divider value, which together determine the transmission period of 1 bit (the elementary time unit or 'etu'). The higher the clock frequency and the smaller the divider, the faster the bit rate. The maximum value of the externally supplied clock is typically 5 MHz. The divider value for transmitting the ATR (answer to reset) as specified by the standard is 372, which results in a transmission rate of 13 440 bit/s with a 5-MHz clock (5 000 000 ÷ 372). However, many smart card operating systems allow the divider to be changed to a different value (typically 64, 32, 16 or even 8) by a performing a PPS (protocol parameter selection) transaction after the ATR.[1] This increases the transmission rate considerably, to as much as 312 500 bit/s (5 000 000 ÷ 16) with a 5-MHz clock and a divider value of 16.

A small marginal comment is in order here. For more than 10 years now, there has been a heated dispute (primarily between German and French card producers) on the

---

[1] See Section 2.3.2

question of whether the T=0 or T=1 transmission protocol yields faster communication, with one camp (French) favouring T=0 and the other camp (German) favouring T=1. Analyses of every imaginable sort have been generated to support each of these positions, and they are repeatedly defended quite vehemently in conferences and standardization committees. However, over the course of time this dispute has come to resemble late-night campfire debates about which beer tastes better. With modern smart card operating systems, the choice of transmission protocol is not a significant technical factor with respect to effective data transmission speed.

To optimize communications, you should focus on certain options at the application level in addition to the transmission protocol parameters. The data provided to the terminal in response to certain commands, such as SELECT, may not actually be needed by the terminal or it may be only rarely needed. In the case of the T=0 transmission protocol, this data is fetched using the GET RESPONSE command. With T=1, the terminal can optionally use the length parameter L_e to request that data be returned in the response to the command.

If the outside world does not need this information, it should not be requested from the smart card, since sending the data costs transmission time and unnecessarily slows down the command sequence. Another approach is to buffer the optional return data for a command in a cache the first time the command is executed. This makes the data available externally without having to fetch it from the smart card each time the command is used. Of course, if you use cached data you must always ensure that the data is valid when it is used.

5.15.3 Commands

Implicit file selection can be used when the files in the smart card are accessed from the application level by read or write commands. With implicit file selection, the short FID of the file is passed as a command parameter and the file is selected automatically during command processing before the file is accessed. This saves the transmission and processing time that would otherwise be necessary for explicit selection using SELECT. However, implicit selection can only be used if the file structure is already known, since otherwise it is not possible to know which read or write command is appropriate (READ BINARY / READ RECORD or UPDATE BINARY / UPDATE RECORD).

If all records of a file must be read, the 'next' option can be used with the READ RECORD command. This allows all records of a file to be read from the beginning to the end without using absolute record addressing. Although this does not provide any direct speed enhancement, it does slightly simplify the application software in the terminal because it eliminates the need to use a record index. In situations where a particular file is accessed repeatedly by read or write commands, it is not necessary to select the file anew for each command.

Of course, the above-mentioned optimization measure does not provide optimum robustness. For instance, if several applications access a smart card using different logical channels,[1] under unfavourable conditions the smart card operating system could lose the context and no longer know which file is currently selected. In this case, it would be

---

[1] See Section 2.3.5

advisable to select the file anew for each read or write command. Whether it is necessary to provide protection against such an extremely rare operating system error at the expense of transaction speed is a question that must be answered individually for each application. It is usually not necessary, but there are nevertheless some mobile telephones that always select each file anew before each access.

If you design an application that uses special, user-generated commands (i.e. commands that do not belong to the standard set of commands supported by the operating system), you should bear two particular considerations in mind. First, atomic processes are very time intensive because of roll-back and roll-forward protection via the EEPROM, so it is advisable to use atomic processes sparingly and very selectively. Second, you should use the functions provided by the programming interfaces of operating system as much as possible instead of yielding to the temptation of implementing most of the functionality in your own code. The functions provided by the interfaces execute much faster because they call optimized program code, which usually runs in the machine language of the target system.

It is advisable to use suitable measuring tools to analyse (profile) the time behaviour of all user-generated commands. In many cases, it takes only a few optimization measures to significantly improve the speed of command processing.

5.15.4   Data and files

With the precursors of modern smart card operating systems, optimization methods fairly close to the hardware level could be used with data stored in nonvolatile memory. This involved taking into account the page size and erased state of the EEPROM when specifying the data to be written. The result was that data was always arranged to occupy an integer number of EEPROM pages so only full EEPROM pages would be written or erased in any given operation. To enable the data to be written quickly, it was also coded in a manner that avoided the need to erase the EEPROM before writing the data in certain cases. It was certainly possible to save up to several hundred milliseconds of processing time for each command by using these measures, depending on the data structure. With modern smart card microcontrollers, the page size and erased state of the EEPROM are no longer externally visible because the EEPROM is subject to memory management and the memory is scrambled, so these measures are not effective. It is thus no longer possible to obtain any benefit from this sort of optimization. In addition, it ties the application code to a particular type of microcontroller and such a high degree of hardware dependence is no longer considered acceptable.

There are several other optimization possibilities that have considerable potential, although they unfortunately affect the design of the application. As a general rule, only data that is absolutely necessary should be stored in the smart card and it should be stored in the most compact format possible. Particularly in the case of character strings, 4-byte UCS characters occupy considerably more memory than single-byte ASCII characters and they significantly increase the transaction time for reading and writing this data.

To the extent allowed by the constraints imposed by the access privileges, you can accelerate data access by combining data into a single file instead of distributing it over several files, even if the latter arrangement would yield a better logical organization. With a single file, only one selection process and one read or write command is necessary to access the data. The hard and fast rule of information technology that data should never be stored redundantly is naturally applicable here also.

There are also optimization measures for data and files that can be implemented without any effect on the application in the smart card. They are used extensively in mobile telephones. For example, mobile telephones read all the necessary data when the phone is switched on and store it in a buffer, so it is available to the internal software of the mobile telephone via an interface without any further access to the smart card. This makes it possible to relocate most of the communication with the smart card to a point in time that is favourable from the user perspective.

A similar measure is to accumulate all the data during a transaction and write it to the smart card as a combined data set at the end of the transaction. However, care must be taken to avoid creating any security gaps if this measure is used, since premature termination of the transaction can result in failure to store all the data in the smart card under certain conditions. Nevertheless, this technique of buffering data in the terminal in order to reduce the transaction time is quite suitable for data that is not related to security. It is also used in practice in many mobile telephones.

Chapter 6

Implementation Patterns

The implementation phase follows the architectural design and specification phase. Implementation consists of transforming an abstract description into source code written in a programming language. A large number of books have been written about the coding process, and they contain innumerable useful suggestions regarding this phase. This chapter focuses on implementation patterns in the smart card context, which has several unusual features. In particular, the methods used for testing smart card applications differ significantly from the methods used in other areas.

6.1   Application Principles

Smart card applications, regardless of whether they are file-based or code-based, should always comply with a set of principles that are generally accepted in the smart card world. This is the only way to ensure that they will work flawlessly with commonly used smart card operating systems and will be reliable, robust, extendable and interoperable. The most important principles for smart card applications are described in the following subsections, with references to other parts of the book that deal with each of these topics in greater detail.

6.1.1   Program code

The program code in smart cards is primarily found in connection with commands, although there is a group of principles that are largely independent of this fact. The most important of these principles are described in the following paragraphs. If Java program code is involved, it is advisable to read Section 6.5 as well.

No secret data   The program code should never contain any secret data such as keys or passwords. This has the advantage that no access to secret information is necessary during software development and programs can be tested by third parties without requiring any special measures to ensure confidentiality. Placing all secret data in files or data objects external to the program code also has the advantage that loading this data can be postponed until one of the final smart card production steps, at which point it can be performed using a fully automatic process.

No debug functions During development, it may be necessary to instrument the program code with supplementary debug outputs and the like. These items must be removed no later than the final test stage, and it must be ensured that program code containing such extensions never reaches any end users of the smart cards. For example, you can use a special ATR to identify a version of the program containing debug functions and test all of the manufactured smart cards during finishing to see whether this ATR is generated. This is a simple, inexpensive measure with a high degree of reliability.

Use standard interfaces The advantage of using standard, generally known interfaces is that it fosters the compatibility of the associated software – and you should not underestimate this advantage. Furthermore, widely used interfaces typically achieve a high level of maturity quite quickly. This is not necessarily true of interfaces that are used only occasionally and proprietary or company-specific interfaces, so they should be used only if necessary.

Confidentiality – even in case of crash The software of a smart card must function robustly under all conditions. If a program crash nevertheless occurs, this must never lead to a breach of the confidentiality of the data stored in the smart card. Generating readouts of the contents of important registers and memory areas when a fault occurs, which is common practice with PC software, must therefore be strictly avoided with smart cards, even if this makes troubleshooting more difficult. Otherwise, the risk of unintentionally sending secret data to the terminal would simply be too large.

No states that survive across reset A cold reset should cause all internal states of the smart card to be reset to their initial values. This ensures that the internal states of the smart card are always the same directly after it is powered up.

Immunity to voltage interruptions Interrupting the supply voltage at any arbitrarily chosen time must not cause any harm to the application or render the smart card unusable. This is one of the most important principles of smart card programming.

Hardware fault handling The smart card should not be permanently deactivated if a hardware fault occurs, such as an error during an EEPROM write, erase or read operation or detection of inadequate quality of the random numbers generated by the smart card. Ideally, the smart card software should include suitable mechanisms for handling detectable hardware faults. This could, for example, consist of switching to a different EEPROM area or reinitializing the random number generator.

Exception handling It must be ensured that all possible exceptions are handled by the software. This requirement has a relatively large number of specific consequences. They include testing the validity of pointers before they are used, testing field limits during accesses, pre-testing for divide by zero, and handling memory request errors, errors during write accesses to nonvolatile memory, file access errors, and of course all forms of type conversion errors.

Defensive programming Smart card software must always be programmed defensively, even if it makes the code more complex in some areas than it would otherwise be. Defensive programming includes ensuring that the software can recognize problems that may be encountered during processing and attempt to initiate an orderly fallback. This can even go as far as replacing obviously incorrect values with allowable values in certain cases so that program execution can continue without premature termination.

6.1.2 Commands

The vast majority of application-specific program code in smart cards is used to implement command functions. Thus, there are several principles specifically relevant to smart card commands that must be observed in addition to the previously mentioned principles for smart card software.

Check data before use After the data of a command has been received by the smart card, the first thing the program code of the command must do is to thoroughly check this data. It is imperative to check aspects such as the length, format, coding, and completeness of the transferred data.[1] A hard and fast principle is that it must never be possible for program execution in a smart card to be disrupted by data received from the outside world. This is a very important principle, particularly with regard to potential attacks.

Meaningful error messages Every error message generated by a smart card, which means all return codes other than '9000' and '61xx', must provide a clear and readily understandable indication of the reason for premature termination of processing of the associated command. However, the data returned to the terminal must never contain any information that could be used for an attack.

High error tolerance Commands should exhibit as much error tolerance as possible relative to the outside world, subject to the condition that it does not allow a potential attacker to acquire additional information. The reason for this is that most terminals pursue a rigid course of action and terminate transactions prematurely if any sort of irregularity occurs. As a result, serious problems can occur in the field if the smart cards are too fussy about details.[2]

No secret commands Secret commands used for error analysis or as backdoors for special cases must always be avoided, since they pose a considerable risk to security and reliability. Such commands are usually not mentioned in any specification and are thus not rigorously tested, so they form a rewarding starting point for many types of attacks.

6.1.3 Data

Besides program code in the form of smart card commands, data is one of the essential components of a smart card application. In many cases, the commands used by the application are limited to those provided by the smart card operating system, so the application effectively comprises only data and associated access privileges.[3]

---

[1] See Section 6.4
[2] See Section 8.4
[3] See Section 3.3.2

No redundant data The data storage capacity of smart cards is relatively small compared with other modern data storage media. If only for this reason, it is a good idea to store all data only once in the smart card instead of storing it redundantly in several places. Another good reason to avoid redundant data is that it is always difficult to maintain the consistency of redundant data. Consequently, you must always ensure that only data that is actually necessary is stored in the smart cards and that each data element is present only once. In cases where it is necessary to access certain data from two different places in the file tree of a smart card, you can utilize the linking mechanism provided by some smart card operating systems.

Testable and manageable data It must be ensured unconditionally that all data elements of a smart card can be tested and managed, as otherwise considerable difficulties can arise in connection with personalization and card administration in the field. This requirement also applies to all secret data. For example, suitable file access privileges must be present for the EFs that hold secret keys, as otherwise these keys cannot be written to the files. For the same administrative reasons, all data of an application must be stored in files or data objects, rather than in parts of the program code.

Manageable secrets People often forget user-specific secrets such as PINs and passwords. Incidentally, this is by far the most common problem with smart cards in the field. In such cases, it is useless to simply reset the error counter. Instead, it must be possible for the user to enter a new secret in the smart card, as otherwise the smart card can no longer be used for its intended purpose. It is equally important to provide some means to reset the error counters used for authentication functions. This makes it possible to use administrative commands to reactivate a smart card that has been blocked by an error counter.

Checksums In some applications, it is important for the smart cards or the outside world to be able to reliably detect corrupted data. If this is necessary, the data concerned should be protected using an error detection code (EDC). A cyclic redundancy checksum (CRC) is typically used for this purpose in the simplest case, although a Reed–Solomon code can be used instead. Using an error correction code (ECC) can lead to difficulties if the accuracy of the data after a correction has been performed cannot be ensured with absolute certainty.

Restrict privileges to the minimum necessary Regardless of whether data is stored in files or data objects, access privileges for all data in the smart card should be restricted as much as possible. Of course, it must be possible to access the data necessary for the intended functions, but all other accesses can be blocked. This is an important prerequisite for ensuring secure storage of data in smart cards.[1]

6.1.4 Security

Secure data storage and program execution are the two essential reasons for using smart cards. With regard to data, security is ensured by the attributes of confidentiality, integrity

---

[1] See Section 5.3

and authenticity, and this security must be an essential criterion in the development process from the requirements analysis stage onwards.

Upgrading an existing system to make it secure is quite difficult, and it can only be done with considerable effort and expense. Naturally, all generally accepted basic principles of cryptology must be observed with all devices for which security is a crucial factor (which of course includes smart cards). These principles are described in a book by Bruce Schneier[1] and other sources. Several interesting examples of mistakes in security systems are described in the publications of Ross Anderson.[2]

Low exposure Smart cards must offer potential attackers as few avenues of attack as possible. Of course, this property is supported by the microcontroller hardware and the smart card operating system, but it must also be borne in mind when developing the application.

This relatively abstract requirement can be illustrated quite clearly by a few specific examples. Appropriate access conditions for read and write access should be specified for all files and data objects present in the smart card. Error counters must always be enabled, and they must have maximum values in the single-digit range.[3] A limit should be placed on the number of possible authentications of the smart card by the outside world (using INTERNAL AUTHENTICATE), and it is important to use strong cryptographic algorithms. Home-brew cryptographic algorithms should be avoided in most cases, as there is a very high probability that they will not be able to withstand attacks.

Verifiable security principles An important consideration is that all security mechanisms that are used, including cryptographic algorithms, authentication procedures and other cryptographic protocols, should be based on verifiable and generally accepted security principles. The security of a system must be able to withstand analysis by neutral third parties as necessary, rather than being based on concealment. This does not necessarily mean that all specifications should be made public, but the security of the system must be able to withstand publication of all of the specifications.

Clarity regarding the use of cryptographic elements The intended objective of the cryptographic elements that are used must be clearly defined starting as early as the requirements phase. The objective can be integrity, confidentiality or authenticity. It is also important to generate a clear specification so that the smart card system can be oriented accordingly.

Protect secrets The secrets stored in a smart card, which means passwords and keys, must be given special protection. Storing this secret information in encrypted form and briefly decrypting it when it is used is a suitable way to achieve this protection, if this is not already implemented in the smart card operating system. As a rule, it is never necessary to grant the outside world read access to secrets stored in a smart card, so such access should always be blocked using suitable access conditions. To strongly curb certain forms of attack, such as brute force, SPA and DPA, it is advisable to use a counter to limit the number of possible authentications if authentication keys are used. It is also

---

[1] See Schneier (1996)
[2] See Anderson and Needham (1995) and Anderson (2001)
[3] See Section 5.11

worthwhile to use different keys for different functions so that a successful attack on one key will not have any direct impact on the other keys.[1]

Protect data during transmission An important principle for smart card applications is that eavesdropping and deleting or inserting data during message transmission must not cause the transaction to be compromised. Consequently, suitable measures must be taken to protect data during transmission, depending on the specific transaction step of the application. Here, you should bear in mind that the simplest solution to this requirement, which is to first add a MAC to the data and then encrypt it, markedly increases the complexity of the application. For this reason, you should always find an acceptable balance between adequate protection and a reasonable level of effort and complexity.

Fallback mechanisms If an attack occurs at any time in a smart card system, the system must be able to recognize it as an attack. It is also essential to have adequate fallback measures available to successfully defend against attacks. For example, there should be mechanisms available for blocking individual smart cards in the system or switching to new keys. A technically sophisticated solution would be to have alternative cryptographic algorithms waiting in reserve in the smart cards so that the system operation could be switched to a different algorithm if necessary.

Of course, such an elaborate fallback mechanism is hardly reasonable in a system with only a few hundred smart cards. With such a system, the cards could be replaced en masse in the event of a successful attack, but this form of card replacement can easily turn into an expensive proposition with somewhat larger systems. Consequently, planning and implementing suitable fallback mechanisms starting as early as the high-level design phase of the smart card system can help avoid saddling the system with significant costs at a much later point in time.

Do not place your trust in third parties A generally accepted basic principle of security-related applications is that, in all cases, you should only rely on functions that are under your own control. A corollary of this principle is that placing trust in security functions provided by third parties can lead to security problems, since these functions are outside your direct control. Even if the specifications are publicly available, it is rarely possible to reliably assess all aspects of a foreign system. This involves more than just examining documents and mechanisms – it also includes aspects that are difficult to assess, such as physical security with regard to generating and storing keys or access to security-related data. It is thus advisable to perform all security-related functions of the smart card system exclusively within an application under your own control.

6.1.5 Application architecture

File-based and code-based applications are always built on top of a smart card operating system. Depending on the type of application, there are several important principles that must be observed in this regard. Failure to do so can lead to considerable difficulties later on. A selection of generally applicable rules and recommendations that do more

---

[1] See Section 5.13

than just put the final polish on smart card applications are described in the following paragraphs.

Defensive design   Generally speaking, smart card applications should always have a defensive architecture because access to the cards in the field is highly limited. This is quite different from the approach commonly taken with PC software, which is characterized by a steady stream of bug fixes and improved versions appearing at monthly intervals. By contrast, after smart cards have been issued it is effectively no longer possible to access them to correct errors or problems. Defensive design imposes the requirement of restrained use of functions provided by the smart card operating system, and it also means that only the functions essential for the application are to be provided.

Robust design   You should always try to achieve seamless integration between the smart card application and the offcard application, as long as this is possible without reducing security. Otherwise, relatively minor specification gaps or software errors can cause the smart cards to not behave as expected in certain situations. This attribute is generally called *robustness*, and it is rather difficult to specify. It is primarily based on many years of experience with the smart card system concerned.[1]

Another aspect of robustness is that smart cards that are withdrawn from a terminal at some randomly chosen time must not become unusable as a result. When a smart card is unexpectedly withdrawn from a terminal, the smart card microcontroller experiences a sudden and unforeseeable power-down. Depending on the command sequence in progress at that time, this can lead to failure to write data to non-volatile memory or cause data to be written incompletely or in a truncated form.

Such corrupted data must never cause a security gap to arise or render the smart card inoperable. This means the smart card must be able to reliably recognize corrupted data. One possible response to such a situation would be to replace the data by correct default values.

A consequence of this requirement is that the application in the smart card should not just have a low number of errors, but instead should be entirely free of errors, even if this is nearly impossible with the present state of software engineering. In addition, the application should be designed to be robust enough to tolerate minor errors on the part of the outside world.

Specified use cases   The data specified for use by the smart card application and the user-developed commands and/or commands provided by the smart card operating system allow an enormous variety of command sequences to be generated. Although these sequences are theoretically acceptable and should not create any technical problems if they are actually used, experience with operational systems has led to the common practice of specifying typical use cases. Some of these use cases may even be included in the application specification and form the basis for tests that simulate actual use and service life measurements. The specified use cases also form a template for developing the offcard application that runs in the terminal or on a PC.

---

[1] Developers often use the humorous designation *telepathic interface* to describe this characteristic – a term that speaks volumes about the desired behaviour and how it can be implemented

Known service life   The non-volatile memory (EEPROM or flash) of a smart card has a limited service life. This relates to the number of write/erase cycles as well as long-term data retention. Typical specification values are a minimum of 500 000 write/erase cycles and 10-year data retention, with each value specified over the full rated temperature and voltage range. These limits on service life must be taken into account in a properly designed smart card application. This requires knowledge of the typical use cases and a statistical estimate of their frequency of use. With this information, the service life can calculated or determined empirically using a test setup. The results obtained from the calculation or test must be evaluated, and they must be taken into account in the application as appropriate.

Limited configuration options   It is difficult to anticipate all possible versions of a smart card application when it is still in the conceptual design stage. Consequently, it is customary to introduce a set of configuration parameters that can be used to enable or disable specific functions before the card is issued or even after the card has been issued. This can lead to difficulties because of the number of possible combinations of individual functions. For instance, if a smart card has eight functions that can be used in any desired combination, this results in $2^8 = 256$ different options. Even in this simple case, it is nearly impossible to test all possible combinations even if the outcomes are limited to positive results, which leads to a level of risk that should not be underestimated.

It is not possible to avoid allowing a certain number of configuration options for any given application, but you should always be aware that every additional configuration parameter increases the risk. Above all, you must avoid a situation in which incomplete analysis of the requirements leads to an enormous number of configuration options. Some applications in actual use have around 1000 configuration parameters, which makes it impossible to test them with a good level of coverage.

Protected configuration parameters   Configuration parameters are an attractive target for potential attackers, because there is a good chance they can be used to disable certain security mechanisms or cause the smart card to malfunction. For this reason, write access to these parameters should only be possible if it is certain that the entity desiring such access is actually authorized to perform the access. To ensure this, smart cards typically test the authenticity of the entity requesting the write access before the access occurs. If a security-critical parameter is involved, you should also consider using secure data transmission for transactions that modify the parameter value, in order to prevent manipulation during the data transfer.

Parameterization via files   The configuration parameters of a smart card can be stored in three different ways: internally via the operating system, in a file, or in an object (in a Java card). Each of these options can be used in smart card applications, but storing parameters in files has distinct advantages. Files can be accessed by the outside world using the standard READ BINARY and UPDATE BINARY commands, and smart card operating systems provide rather sophisticated rule-based access mechanisms for files. If the other two options are used, all this has to be reproduced, which entails additional cost and effort. The optimum form of file-based storage can be achieved by storing parameters in a file with TLV-structured records, since GET DATA and PUT DATA can then be used to address each parameter individually.

Log files   Part of the memory allocated to a smart card application should always be reserved for log files[1] used to record major application events. These log files can be read from time to time by the background system to check for unusual events or monitor system consistency. You should always try to make the processes that write to log files robust enough to enable the log files to be read for a post-mortem analysis if an application becomes blocked or inoperable as a result of some event.

Avoid total blocking   It is sometimes necessary to block an application or even an entire smart card. If reversible blocking is not sufficient, it is advisable to block only the specific function instead of disabling the entire smart card when irreversible blocking is used. The personal data of the user (such as a telephone list), administrative data (such as log data), and important application data (such as a purse balance) should still be readable even after irreversible blocking. Otherwise it will no longer be possible to recover important data or analyse the blocked smart card. The DEACTIVATE FILE command is thus much more suitable for blocking directories and files than TERMINATE DF and TERMINATE EF, since the latter commands initiate irreversible blocking with no possibility of subsequent access.

Traceability   To ensure that smart cards can be identified uniquely during their entire life cycle, each card should always be assigned a unique, individual card number (ICCID – integrated chip card identifier). It must be possible to read the card number externally in every state without any restrictions. The card number is the unique identification code of the card, and it can be used as a reference for a broad range of data and activities, starting with production of the card and extending to its destruction. For example, the card number is often used as a reference for querying the database of card finishing log data, which contains information about the production machines that were used, the production site, initialization, personalization, and the date and time of each of the finishing steps.

Destruction of units of value   If a smart card application contains an electronic purse, all unplanned events – such as premature termination, attempted manipulation and the like – must always lead at most to destruction of units of value and never to creation of units of value.

Encapsulation of applications   All files of a smart card application should always be encapsulated (grouped) in a DF, or in several DFs in the case of a relatively large application. The advantage of this is that an encapsulated application can easily be migrated to a multiapplication card. All that is required for this is to insert the appropriate DF (containing the application files and the access privileges in EF_{ARR}) into the file tree of the new smart card. This cannot lead to duplication of FIDs, since all the files are located in an application DF with a unique AID.

Risk-aware use of memory management   The following recommendation will presumably be regarded as nonsense by all producers of smart card operating systems. Nevertheless, it should not be omitted from the list of application principles. Until around the mid-1990s, most smart cards did not have any functions for creating new files or

---

[1] See Section 5.4

deleting existing files, since these operations are rather complicated from a software perspective and are a rich source of potential errors. The situation has changed completely in the meantime, with the result that modern smart cards have file management capabilities that are fully comparable with those of PCs.

However, as an application developer you should always be aware that file management commands such as CREATE FILE, DELETE FILE, LOAD and INSTALL entail extensive operations in the memory management system of the smart card. It is thus possible for an inconsistency to occur in memory management, despite all security measures, if an unforeseen power-down occurs or there is a fault in the nonvolatile memory of the smart card. If the operating system detects the inconsistency, it may permanently disable the smart card. For this reason, it is advisable to use commands for managing files and applets warily instead of employing them on every possible occasion 'just for the fun of it'.

6.1.6 System

The ultimate objective of every smart card application development project is to produce a complete, properly operating system – and the smart cards are only a small part of the overall system. Nevertheless, the smart cards are one of the most important parts of the system, since they contain all the confidential information of the system users. There are several related aspects that you should pay attention to from a system perspective. They are described in the following paragraphs.

Awareness of the types of limits It is worthwhile to give detailed consideration to the limits of the system, and naturally also to the limits of the smart cards, when specifying the requirements and elaborating the specifications. These limits can be roughly classified into three types. The first is physical limits, which are set by immutable natural laws and cannot be got around by even the most sophisticated technology. These limits are often indicated by the term 'must' in standards. A good example of a physical limit is that the maximum data transmission rate of an I/O conductor is limited by its electrical capacitance, among other things, since the current supplied by the driver must be restricted to a certain level. This sort of limit can only be circumvented by using a completely different technology, such as optical data transmission.

The second type of limit, which is not as hard as the first type, is imposed by the technology or standards that are used. Limits of this type can be extended significantly by new technology or new methods. For example, the maximum data transmission rate between the terminal and the smart card corresponds to the applied clock frequency, but this limit is only imposed by the T=1 transmission protocol. A significantly higher maximum data rate is possible if the USB protocol is used.

The third type of limit, and the one that changes the most rapidly, is a limit imposed by social and cultural constraints, which are specified by laws, regulations and ordinances. Probably the best example here is data protection legislation, which can be rendered ineffective in broad measure within only a few weeks in response to suitable events.

As changes to limits can impact the functionality of the smart cards and the system, it is highly important to be aware of the degree of flexibility of each of the limits. For instance, a smart card system designed for unconditional compliance with data protection legislation would experience considerable difficulties if it suddenly became necessary to grant government security authorities access to the data in the smart cards. The hardness of the limits depends on the individual limits, and this must be reflected in the system design. This is a prerequisite for being able to respond flexibly to changed conditions without having to make massive modifications to the overall system. Of course, this directly contradicts the requirement that smart card applications should have as few parameters as possible. For exactly this reason, it is a good example of how difficult it is to develop good smart applications that are viable in actual use.

Judicious use of remote maintenance   In smart card systems in which the cards have regular contact with a background system, the understandable desire to perform administrative activities during these contacts arises almost inevitably. This can be done without creating any major technical difficulties, but it does create a logistical problem because there may be several card versions in the field after a few years instead of only one, depending on external circumstances.[1] In such cases, it is important to determine the technical parameters of the smart cards concerned before performing each administrative activity and take them into account accordingly.

For the same reason, it is always advisable to use remote operations with great care – especially administrative operations – and limit them to exceptional cases. This is because, in the worst case, smart cards can be rendered inoperable by poorly conceived remote administration or maintenance activities. Experience shows that users respond indignantly in such cases and take a very dim view of the system operator.

Damage limitation   A principle that is applicable to all smart card systems is that even if the secrets of a few smart cards are ferreted out, this must not cause the entire system to be discredited. This implies using a suitable key management scheme[2] with no master keys of the system stored in the smart cards. You can depart from this principle in favour of a simpler scheme if the system is fairly small, but in all other systems the smart cards should only use keys derived from a master key. This ensures that it will not be necessary to replace all the smart cards in the system if one of the cards is successfully attacked.

System consistency   One of the essential tasks of a background system is to monitor the consistency of the system, either continually or at reasonably frequent intervals. This is indispensable for the proper operation of the system, since it is the only way to determine whether there are problems in the system – such as execution of a successful attack or a malfunction in one of the components. In the case of an electronic purse system, consistency checking is typically performed by comparing the sum of the load amounts for each card with the sum of the payments for the same card. The two sums should match over a moderate time interval, since otherwise it is likely that someone has succeeded in illicitly increasing the credit balance in his smart card. Another commonly used practice is to maintain a shadow account that reflects the purse balances of the smart cards.[3]

---

[1] See Section 7.2
[2] See Section 5.13
[3] See Section 7.3

Meaningful approaches to checking system consistency depend strongly on the functionality of the smart card applications concerned and must be specified on a case-by-case basis. In this connection, you must also ensure that you do not violate the principles of data protection.[1] In any case, it is easier to monitor consistency in online systems because all components can be accessed directly. Nevertheless, accurate consistency checking is more important in offline systems because operational problems are more difficult to detect in such systems and can only be seen some time after their occurrence.

Smart cards are the weakest component   In a typical smart card system, the net worth of the cards in the field can easily be regarded as considerable. Nevertheless, there is evidently a preference for making any necessary modifications to the system in the smart cards instead of in other components, regardless of the underlying reason for the modifications, as can be seen again and again in actual practice. Although this behaviour is difficult to justify logically in many cases, it is a fact and can certainly be taken into account when planning an application development.

6.2   Testing

Software in embedded applications – which includes smart card software – is subject to stringent quality requirements. According to ISO 8402, quality is defined as the totality of features and characteristics of a product or service that bear on its ability to satisfy the stated or implied needs. The two most important attributes of quality are robustness and freedom from defects and errors.[2] As these attributes cannot be achieved solely by the development process, testing using adequate and suitable tests is also necessary.

The purpose of testing is always to find defects and errors, and the entire testing process must be oriented towards this purpose. Tests are always random samples and can never be complete, since it is never possible to fully work through all possible permutations.

The following paragraphs provide some suggestions and recommendations for testing smart card applications. Good treatments of testing as a development activity can be found in books by Peter Liggesmeyer[3] and Andreas Spillner.[4] See Section 7.1 for information about tests related to card finishing.

As an incentive to encourage thorough testing, it can be remarked here that you can reasonably assume that anything that is not tested will later turn out to not work properly. Incidentally, you must always avoid a situation in which the people responsible for developing the application software assume that any defects in the software will be eliminated by testing at the end of the development process. This is a mistaken attitude – the software must have an adequate level of quality before it is handed over to the test team.

---

[1] See Section 4.1

[2] For many years now, the licence agreements of a major American software producer have included a clause stating that given the current state of software development, it is not possible to produce software that is totally free of defects and errors. For this reason, users must be satisfied with what they get. Although this business practice of a few firms in a monopoly position may lie just within the limits of what is acceptable in the case of PC software, users of embedded software expect their software to be free of defects instead of 'low-defect'. This means the software must be stable under practically all conditions

[3] See Liggesmeyer (2002)

[4] See Spillner and Linz (2003)

Test types Many types of tests are used in the smart card world, depending on the specific phase of the life cycle. Module tests are used to test individual software modules. This generally requires creating a test environment, since modules cannot run by themselves. Integration tests, which are built on top of module tests, involve integrating a group of modules according to their intended function in the system and testing them together. These tests focus primarily on the combined functions of the modules and the interfaces between the individual modules. System tests occupy the highest position in the hierarchy. Their purpose is to test the entire smart card system comprising all the software modules.

By contrast, regression tests are used to check for undesirable side effects after the program code or application has been modified or updated.

A formal acceptance test is normally conducted before delivery to the customer. Its purpose is to enable the customer to verify that their requirements have been fulfilled and the desired level of quality is present. Compatibility tests are used to verify mutual compatibility of the smart cards and their compatibility with the rest of the system.

This amounts to a brief description of the types of tests typically used with smart cards. However, these tests can vary considerably depending on the specific perspective. For example, testing at the system level tests the entire system – consisting of the background system, the network, the terminals and the smart cards – instead of just the smart cards, which are one of several components of the overall system.

Test cases A test case defines a specific test in terms of its purpose, prerequisites, inputs, test procedure, and expected results. Test suites are always composed of several test cases.

Test scenarios Commonly used test scenarios can be arranged hierarchically according to their order of precedence. The first level consists of command tests. They test individual smart card commands using a separate test for each command, without giving any consideration to how they relate to other commands. For example, a command test could be used to test the READ BINARY command with all its parameters. The next level is command sequence tests, which encompass two or more commands and their interactions. An example of a command sequence test would be a test of the GET CHALLENGE and EXTERNAL AUTHENTICATE commands, which can be used to authenticate the outside world relative to a smart card.

The next level in the scenario hierarchy consists of tests of typical use cases, which consist of several different commands and command sequences. Some typical use cases in an electronic purse system are payment, loading and cancellation, and the complete start-up sequence of an application is another typical use case. Incidentally, a common, but often overlooked, use case is initialization and personalization of the smart cards.

The above-mentioned test scenarios form the basis for testing the life expectancy of the smart cards. This testing is based on the specified use cases and their probability of occurrence. The primary objective of service life testing is to determine when the first EEPROM errors can be expected. Endurance tests are usually used for this purpose. The service life can only be determined reliably using empirical methods – mathematical methods are not sufficiently trustworthy.

This bottom-up approach to test scenarios, which is commonly used in the industry, can be used to test smart card applications systematically by starting with individual commands and finishing with the entire life cycle. The required cost and effort also lie within an acceptable range, since the individual test modules, which are initially used for command tests, can be reused repeatedly. Of course, strong configurability of the modules is an important factor for reusability.

Test specifications   Each test case and its significant parameters must be specified in order to document the tests. This is traditionally done in the form of separate documents containing all the necessary information. However, a natural consequence of this approach is that the actual tests and their specification documents diverge after several revisions. A modern documentation method is to incorporate the specification directly in the source code of each test. A standard extraction tool (such as Doxygen) can then be used to generate the specification directly from the source code, in the form of a hypertext document. This method significantly increases the likelihood that the test specification will fully describe the source code of the test. In addition, the total cost is significantly less than the cost of generating and maintaining separate test specification documents.

An indispensable element of a test specification is a capsule summary of the defined test cases, since this makes checking for gaps in the test cases significantly easier. If you use a hypertext document for the specification, you can also insert cross-references to access the test cases and the associated source code directly from the summary.

Test development   Software for smart cards and the associated tests should always be developed using the 'four eyes' principle, which means that two people are always involved. This ensures that the implemented functions are checked by two independent persons, and it also promotes early discovery of portions of the specification that are subject to interpretation.

Ideally, the functionality of the smart card software should be checked by a separate, independent test team. If the cost of such a team is too high in a particular case, the functionality can be checked by a developer instead. However, this should be someone other than the developer who programmed the software that is to be checked.

Test structure   All individual smart card tests have the same four-stage structure (Table 6.1). The first stage consists of creating the conditions in the test environment necessary for performing the test. The next stage is performing the test, which is followed by comparing the test results with the expected results. Finally, the test environment is restored to its original state so that the conditions in the smart card following completion of the test are the same as those before the start of the test. This must be done even if the test result does not match the expected result. Otherwise, it is not possible to perform multiple tests in series independently of the results of the individual tests.

The generic process described above can be illustrated using an example. Assume that the purpose of the test is to check whether a file with a transparent structure can be read using the READ BINARY command. To produce the right conditions for the test, a transparent file must be first created in a smart card. This file is then populated with data so that the data read from the file when a READ BINARY command is executed can be checked. During the actual test, data must be read from the card using several different data lengths and offsets. As the data content of the file is known, the test program can

Table 6.1 Basic structure of an automated test for smart cards

Step	Action	Explanation
1.	Test build-up	Create the necessary conditions for performing the test
2.	Test execution	Perform the test
3.	Check against the expected result	Analyse the test result
4.	Test build-down	Restore the conditions present before the test build-up

check the received data against the expected data. If they match, the command has been executed correctly. If there are differences, an error has occurred in the smart card. Regardless of the result, the data in the smart card must be deleted after the command has been tested in order to restore the original initial conditions. In practice, tests of this type must be performed using a large variety of parameters and several different files ranging in size from zero to the maximum value.

Test modularization   Tests for smart cards must be designed so that they can run automatically without any manual intervention. It is common practice to construct tests in modular, mutually independent form. These modules can be chained together for specific test tasks and run automatically one after the other. This is illustrated graphically in Figure 6.1

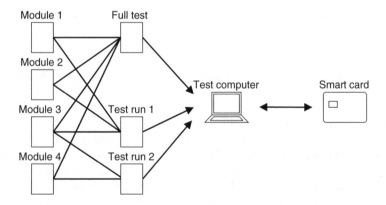

Figure 6.1   Using test modules as components of scalable test runs. This makes it possible to perform time-optimized test runs corresponding to actual requirements

This structure makes the tests scalable, so the tester can run a single test for a relatively minor modification to the smart card software or the entire suite of tests for a major revision. This approach significantly reduces the time required to perform testing, since it can be restricted to the test modules necessary for the specific purpose of the test. Development lead time can thus be reduced significantly. The extent of the time savings increases with the scope of the tests. To cite one example, a full set of tests for a smart card operating system typically encompasses around 50 000 test cases with a total of more than 300 000 application protocol data units (APDUs) and an execution time of approximately 25 h.

Modularly structured tests are also easier to configure for concurrent execution, so several test modules can be run in parallel on a set of test computers when a large number of test cases are present. The time required to perform the tests can thus be reduced in direct proportion to the number of test computers used.

Test automation It is relatively easy to automate tests for smart cards because there is only one interface, which can be completely monitored and controlled by the terminal. This is illustrated in Figure 6.2. Manual operations, which are difficult to automate, are not required with smart cards (unlike the situation with mobile phones, for example). As a result, fully automated testing of smart card applications without human intervention became established relatively early. This has the advantage that a large number of tests can be performed in a short time. However, attributes such as modularity, restartability and logging are essential for automated testing.

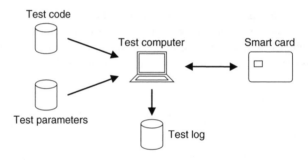

Figure 6.2 Components and communications in a typical fully automated smart card test setup

Test logging Ensuring the traceability of the tests that are performed requires more than just using unique version numbers. It is also necessary to log the tests at the APDU level. For this purpose, all command APDUs and response APDUs transmitted during the test, the execution time, and related information about the test case are stored in a log file. The data in the log file can be formatted as pure ASCII in the simplest case, or a modern format such as HTML or XML can be used. A sample log entry is shown in Table 6.2.

If the test is very extensive and processes several thousand APDUs, it is difficult to determine, after completion of the test, whether the entire test ran without any errors. For this reason, it is advisable to check the received response APDUs against the corresponding expected results directly during the test. The results obtained this way can be summarized at the end of the log file for rapid evaluation. This makes it possible reach a firm conclusion regarding the overall result of the test after all the individual tests have been completed without a time-consuming analysis of the log file.

Table 6.2 Sample record of a log file containing test results. If the file contains HTML-formatted data, hyperlinks to other documents can also be entered

...	
Test module:	READ BINARY
Test case:	Check maximum file length of 10 in test file 1
Test specification:	Hyperlink to the test specification
Command APDU:	'00 B0 00 00 0A'
Response APDU:	Data: '01 02 03 04 05 06 07 08 09 10'
	Return code: '90 00'
Execution time:	1.697 ms
Result:	OK
...	

Restarting It is important to ensure that a failed test never leads to a 'hung test', which means premature termination of a series of modular tests. As relatively large test suites are often run unattended overnight, the test modules must be able to start again without human intervention after an error has been detected. Otherwise, termination of any one of the tests would prevent running the subsequent tests in the series. If this happens with automated tests running overnight, you may find that only a small number of the full set of tests have actually been run.

Test quality As tests can never fully cover everything, it is difficult to define a completion criterion for test development. The simplest approach is based on the cost and effort of developing the tests. For typical smart card applications, you can use the rule of thumb that the same amount of effort should be spent on test development as on product development. A somewhat more refined approach is to try to achieve full requirements coverage with the tests. This is possible without extraordinary effort even with complicated applications. In addition, you should ensure that the tests encompass the positive outcomes for all use cases and the associated negative outcomes that can typically be expected. Equivalence classes are commonly used to test value ranges.

Finally, you should measure the instruction coverage and branch coverage of all available tests. The minimum level of instruction coverage should be in the 95% range, and the branch coverage should be in the 80% range.[1] If these levels are not achieved, or if you find even larger contiguous gaps during coverage analysis, it is advisable to extend the tests accordingly.

Optionally, you can also schedule a certain amount of time for freely defined testing by experienced testers after completion of the specified tests. These testers can check the finished application using tests based on their accumulated personal experience. These tests must also be logged in the interest of subsequent traceability.

If you have an opportunity to test the smart cards in a real system in addition to the previously described tests, you should most certainly use it. It hardly needs to be emphasized

---

[1] See Section 8.2.5

that scenarios involving test computers, no matter how good they may be, can never replace testing under real conditions.

The procedure described here usually produces adequate test quality and yields smart card applications that have a relatively low number of defects in normal use cases. Of course, significantly more sophisticated test strategies are also possible. The limit is always set by the associated cost, which must be justifiable with respect to the application concerned.

6.3   User–Terminal Interface

The subject of user interfaces is immense, and it has many different facets. Here, the essential aspects of the interface between the terminal and the user are described in summary form, independent of the specific types of terminals or applications.

User secrets   The terminal must use secrets entered by the user immediately and then delete them. This applies primarily to PIN codes, which must never be stored for any other use. The reason for this is that in many cases entry of a secret by a user represents an explicit expression of intent, and the terminal must never carry out a substitute performance of this expression of intent.

The problem of intermediate storage of PIN codes occurs repeatedly with mobile telephones in particular. Some time ago, there was an operating system for mobile telephones that stored the PIN code in unencrypted form in the registry. A thief or someone who found a lost telephone could thus read the stored PIN and then enter it into the SIM for verification. After this, the 'new' user would be able to make unauthorized calls or data transfers at the subscriber's expense. Situations like this must never occur with terminals.

Entering secrets   It must be possible to enter secrets at terminals such that they cannot be detected by third parties. This requires suitable visual screening of the keypad in the case of terminals used in public places. Furthermore, the entered characters must not be shown as plain text on the display of the terminal. Only an indication that the characters have been entered should be shown. Naturally, surveillance cameras must be oriented so that only the user is recorded, but not the keypad entry.

Response time   The transaction time after a contact smart card is inserted in a terminal should not exceed 1 second. In case of actions that are important from the user's perspective, such as generating a digital signature for a document, the transaction time can be somewhat longer in exceptional cases. However, the time between inserting the card and receiving a withdrawal request should never be more than 5 second.

In case of contactless smart cards, which are primarily used on account of their relatively short transaction times, the interval should not exceed 200 ms. Otherwise it will not be possible to complete a full transaction when a user waves a smart card past the antenna of a terminal with a swing of the arm.

If the transaction times are longer than the stated values, users will often receive the impression that something is wrong with the terminal and will either pull the card out of the terminal or wave the card impatiently back and forth in front of the terminal antenna (in case of a contactless card). This reduces user acceptance, and stability of the system can easily be affected if a large number of incomplete transactions occur owing to hectic user behaviour.

Processes It gives users a feeling of confidence if the processes for the various functions of the terminal are largely similar. This includes the sequence of activities for entering data and the terminology used in the user interface. A well-conceived, uniform system of user prompts makes a significant contribution to good user acceptance and consequently leads to fewer aborted transactions and cancellations.

Short-circuit immunity Terminals for contact smart cards should always be immune to persistent short circuits. This will reliably prevent them from being permanently damaged if someone inserts a piece of sheet metal instead of a smart card.

An amusing example of what can happen here arose from a special request by a system operator in the Arabian world, who wanted to have a smart card producer supply gold-plated cards. The producer finally succeeded in making gold-plated cards after considerable technical effort, but a fatal problem arose when they were put into service – despite fact that the card producer left a gap in the plating around the chip module to prevent short circuits from occurring in this area. The problem arose from the fact that the terminals were not immune to short circuits, so they were rendered electrically defective each time a smart card was inserted the wrong way round. Although this was a fairly unusual situation, since gold-plated cards are rather rare, it provides a striking example of the types of problems that can occur if the terminals are not immune to short circuits.

Card detection Mechanical detection of card insertion should be used instead of alternative methods such as a light barrier. A light barrier will not work properly if a transparent card is used, which is not all that uncommon. Problems can occur even if the card is not transparent to visible light, since a smart card that appears to be opaque to the human eye can still be transparent in the infrared region. Many light-barrier detectors use infrared light, so a terminal using such a detector will not recognize that a card has been inserted if the card body is transparent to infrared light.

6.4 Smart Card Commands

Code-based applications[1] are composed of not only files or data objects with associated access conditions, but also smart card commands that perform a wide variety of functions. These commands are typically developed using the Java programming language, and the resulting code is interpreted in the smart card.

First of all, there is a general rule that applies to all smart card commands. It must never be possible under any circumstances to use data passed with a command to cause program execution to crash or pry secrets out of a smart card. You must keep this rule in the back of your mind during every step of the specification generation and application

---

[1] See Section 3.3.3

implementation process. A consequence of this rule is that the majority of the command processing code is devoted to checking the input values.

6.4.1 Command structure

Almost all smart card commands have a similar basic structure, even if they implement totally different functions. Table 6.3 shows the general sequence of actions for processing a command. The command interpreter of the smart card operating system first analyses the class byte (CLA) and instruction byte (INS) of the received APDU[1] and then calls the program code of the command based on this analysis.

Table 6.3   Summary of the basic steps for processing a command in a smart card

Step	Activity
1.	Receive command APDU
2.	Check the APDU header against technical criteria (e.g. that the values of P1, P2, L_c and L_e are within their allowed ranges)
3.	Check the APDU body against technical criteria (e.g. that the data objects needed for processing are present)
4.	Check for compliance with all necessary security conditions (e.g. that file access privileges are present)
5.	Check the command APDU against the content-related conditions (e.g. that the offset lies within the file boundaries in case of file access)
6.	Execute the code that implements the command function
7.	Generate the response APDU
8.	Transmit the response APDU

From an abstract viewpoint, the first step in processing a smart card command is to check and format the input data. The next step is to execute the code of the command function, and the final step is to format the response data.

In the actual program code, the first step is to check the command header against technical criteria. The header typically contains information related to the various command options. The header is usually checked against the minimum and maximum permissible values of P1, P2, L_c and L_e, and/or it is checked for allowed or prohibited values. If the command has a body, the body is checked in a similar manner. In the interest of maintaining a simple structure, the checks usually start with the first byte of the input data and finish with the final byte. However, a different sequence is certainly possible, since the order of the return codes is normally not specified. If a value that is not allowed is detected in the header or body of the command, command processing is terminated and a suitable return code is sent back to the terminal to inform why processing was terminated. It is advisable to select return codes that are as meaningful as possible, since this considerably simplifies troubleshooting.

The third step is to check the states and security conditions that must be satisfied before the command can be executed. An example of state checking would be to verify that

---

[1] See Section 2.3.3

an exchange of random numbers, which is a prerequisite for a subsequent authentication command, has already taken place. Checking the security conditions typically amounts to comparing the security state necessary for accessing a file or data object with the currently attained security state. For example, if read access to a particular file requires successful PIN verification, a check is made here to see whether this has occurred. Access condition checking can also be quite elaborate, particularly when rule-based access conditions are used.[1] In the interest of robustness, a check should always be made before every file access to verify that the file actually exists.

The command processing conditions related to the content of the command are checked in the fourth step of command processing. For example, if the command pertains to reading a transparent file (READ BINARY), this step includes verifying that the command will not cause more data to be read than is actually present in the file and that the offset value passed with the command will not cause data extending past the end of the file to be addressed during the read operation.

At this point, all prerequisites for executing the actual function have been checked prior to execution. However, the next step before the function is actually executed is to format the input data properly. After this, the program code of the command function is executed. This code must be formulated as robustly as possible. For example, every possible exception (such as division by zero, underflow or overflow of a field, and the like) should be handled completely and should cause a meaningful error message to be returned to the terminal. After the function code has been executed, the output data is formatted as necessary.

If the core of the command has been executed correctly, the final step consists of generating a suitable response with a body appropriate to the command. The program flow then returns to the command interpreter, which transmits the response APDU to the terminal and waits for a new command.

6.4.2 Interruption of commands

In the design and implementation of commands, you should always bear in mind that the supply voltage can be interrupted at any arbitrary point in time. In the case of write operations to nonvolatile memory (EEPROM or flash), this can result in incomplete writing of data to a file or data object. If this happens, the data content is undefined. This possibility must be taken into account in the command. Truncation of data due to interrupted write operations can be reliably avoided by using atomic processes.[2] However, this comes at the expense of increased command execution time, which is undesirable in many cases. As a rule, execution times of commands should be less than 1 second, including the total data transfer time. From this, it can be concluded that the elapsed time available for processing the command is typically several hundred milliseconds exclusive of the data transfer time.

Atomic processes should thus only be used when consistency of the written data must be ensured reliably. Data for which this is not necessary should be written in the conventional manner, which is faster. However, the fact that this data can be corrupt must

---

[1] See Section 2.1.6
[2] See Section 2.4.2

be taken into account when it is read. In this regard, commands must always be implemented such that corrupted data in nonvolatile memory can never cause a crash or result
in a deadlock situation during processing of a command sequence. It is also advisable
to check all data read from nonvolatile memory for plausibility or add checksums to
the data. If the data read from memory is corrupt, an error message can be sent to the
terminal or specified default values can be used instead. The latter solution is distinctly
more robust.

6.4.3 Command coding

The header usually contains the general options and parameters of the command, while
the body contains the associated data. This can be illustrated using the VERIFY command (as defined in ISO/IEC 78126-4), which is used to verify a PIN code. The header
parameters of this command identify (among other things) which PIN of the smart card
is to be used for the verification, while the PIN to be verified is transferred in the body.

To avoid potential confusion, it is also advisable for each command to have only one
specific case.[1] It goes without saying that all commands must be independent of any
particular data transmission protocol.

When coding the class and instruction bytes of a command, you should take care to avoid
any conflict with codes used in large international applications such as SIM cards,[2] USIM
cards[3], EMV cards,[4] as well as codes defined in the ISO/IEC 7816 family of standards.[5]
If an application with new commands is intended to be used in a multiapplication card,
the codes defined in the above-mentioned specifications must not be used under any
circumstances, as otherwise there is a high probability of conflict with other applications.
If it is necessary to introduce a new command instead of using an established command,
it is always advisable to base it on existing command specifications. An example of
a typical command specification, including the associated response and possible return
codes, is described in Section 4.5.2.3.

The input and output data of a command should be defined in a manner that is as future-
proof as possible, which means providing sufficient possibilities for later extensions.
For this reason, unused bits in data bytes should be marked as 'reserved for future use'
(RFU) instead of being coded with fixed values. If the body consists of more than one
data item, it should be coded using TLV structures so that it can be expanded to include
other data if necessary. This circumvents the need to define fixed sizes in the command
specification and thus creates considerable manoeuvring room for future extensions.

6.4.4 Parameterization

There are many cases in which it is necessary to parameterize commands or command
sequences, including across session boundaries. For example, this could involve config-
uring the input data of a command with a specific set of values defined for a particular

---

[1] See Section 2.3.3
[2] See TS 51.011 (2003)
[3] See TS 31.102 (2003) and TS 102.222 (2005)
[4] See EMV Book 3 (2004)
[5] See ISO/IEC 7816-4 (2005), ISO/IEC 7816-8 (2004), and ISO/IEC 7816-9 (2004)

purpose while still allowing the configuration to be modified if necessary. To ensure that the values cannot be modified at will, the configuration must be stored permanently in the nonvolatile memory of the smart card.

Two basic approaches to implementing such capability are used in the Java Card environment. One approach is to store the information as a data object managed by the Java runtime environment, and the other is to store the information in a file. Although both of these options can be implemented at comparable cost, they have significant differences with regard to practical considerations. Storing the information in data objects is straightforward from the perspective of a Java applet, since access to the data is direct and simple. However, other operations (personalization, testing, and modifying the information from outside the smart card) are complicated because the information can only be accessed using Java mechanisms. This requires a complete management interface in the relevant Java applet in the smart card. Consequently, it is common practice to store parameter data in files instead of in Java objects. These files can be accessed using established mechanisms (SELECT, READ, UPDATE, etc.), and read and write access to the file can be controlled using the advanced file access conditions of smart cards. Consequently, file-based parameterization should normally be the preferred solution.

6.4.5   Test commands

During application development, it is unfortunately sometimes necessary to introduce one or more commands that are used exclusively for testing. These may be commands that transfer test data to specific modules in the smart card or commands that can optionally read out data without being subject to access restrictions. There is little reason to object to such commands when the application is in the development phase, although they represent a certain security risk. However, if a smart card with such commands were to find its way into normal use, the commands could be used to read out the secrets of the smart card, which at minimum could be used to produce clones of the card. For this reason, it is extremely important to ensure that all test commands are not only disabled on completion of the development phase, but also deleted from the source code. As a supplementary security measure, smart cards should be checked for the absence of test commands at the end of the production process. This creates an additional barrier for reliable prevention of issuing smart cards containing test commands.

6.4.6   Secret commands

Secret (i.e. undocumented) commands that perform various functions while circumventing defined security measures make the news with a certain regularity in the PC world. Such commands are often used to create secret paths for developers so that they can access specific functions for maintenance or espionage purposes even after the product has been delivered. As these commands are undocumented, they are not included in security assessments. Furthermore, they are typically not tested, with the result that they usually contain defects. This situation can be used as a basis for attacks.

Such trap doors must be avoided without fail in smart cards, because the available commands and their allowable parameters can be ferreted out relatively quickly using simple equipment, even without access to the documentation, by simple trial and error. Secret commands will thus be discovered rapidly, and their functions can also be determined with a reasonable amount of effort. Secret commands and undocumented commands thus represent a major security gap, which must avoided unconditionally in the interest of the overall security of the system.

6.5   Java Card

Java is the predominant programming language for applications based on executable code. However, the hardware of smart card microcontrollers has significantly less processing power than PC hardware, so a subset of the well-known Java for PCs has been defined by the Java Card specifications.[1] These specifications provide the most important functions for smart card applications and enable them to be used with acceptable execution speed and memory space requirements by eliminating the language elements and packages that are not important for these applications, such as floating-point numbers, threading, GUI, database functions, and server functions.

Three additional mechanisms not present in Java for PCs were added to fulfil the specific requirements of smart card applications. They provide the following supplementary capabilities:

- define persistent and transient objects

- specify atomic processes

- specify data exchanges between applets and compartmentalization of applets

A detailed introduction and guide to programming applications in Java can be found in Zhiqun Chen.[2] Good overviews of the essential aspects of professional programming are provided by Steve McConnell[3] and Andrew Hunt.[4]

With relatively large Java applications, there is still a risk that the application may run too slowly or require too much memory. If this is only noticed towards the end of the development process, a supplementary optimization stage will be necessary. However, it is better to take the specific aspects of Java Card into account from the start of the software development process. Just as with quality, it is rather difficult to significantly increase the speed or reduce the size of a finished program.

The storage requirements of an applet are divided into storage space for the executable program code and storage space for the data. The program code is always located in the nonvolatile memory of the smart card, which means EEPROM or flash memory. The nonvolatile memory capacity of modern smart cards that support executable program code lies in the range of 16–512 KB. At least at the upper end of this range, the available capacity is adequate for all applications. In unusual cases and with applications involving a large number of issued cards, the program code of a Java applet is sometimes even

---

[1] See JCVMS and JCRES
[2] See Chen (2000)
[3] See McConnell (2002)
[4] See Hunt and Thomas (1999)

integrated into the ROM. However, this requires generating a special version of the smart card operating system with its own ROM mask.

The data of the applet can be located in nonvolatile memory (EEPROM or flash) or in volatile memory (RAM). This is essentially determined by the application requirements, although it can be influenced by the designer. However, RAM is a very scarce resource in smart cards. The amount of RAM available to a Java applet is typically somewhere between a few hundred bytes and several thousand bytes. The stack of the Java VM also needs RAM, which it uses to store return addresses, parameters passed to methods and returned by methods, and local variables. The memory allocation is shown graphically in Figure 6.3.

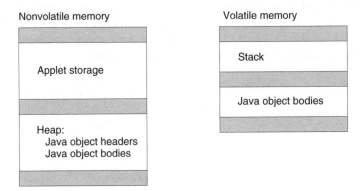

Figure 6.3 Program and data storage in the nonvolatile and volatile memories of a Java card. Nonvolatile memory is called persistent memory in Java terminology, while volatile memory is called transient memory

A Java object consists of a header and a body. The header, which contains all the administrative data of the object, is located in an area of the nonvolatile memory called the *heap*. The body, which contains the actual data of the object, can also be located in the heap and thus in nonvolatile memory. However, several methods are available (such as 'makeTransient()') for storing the body in volatile memory.

Optimization measures always impact the structure of the software in some way. In most cases, there is also a conflict between the objectives of speed optimization and storage space minimization, since these two characteristics are usually mutually contradictory. The art of professional software development consists of achieving a satisfactory balance between speed and memory usage on the one hand and an acceptable structure and robustness on the other hand. However, you should never try to meet this challenge by working in the same way as in the early days of software development. Although it was necessary to comply with the hard constraints of low processing power and limited memory in those days as well, the development processes often bore more resemblance to a handicraft than to an engineering discipline.

Another rather general comment about software development in a microcontroller environment is warranted here. In this sort of development, it is always useful to occasionally examine the machine code generated from the high-level source code and analyse the time and space demands of the commands and individual routines. These analysis functions are provided by commonly used development tools. The information obtained from these analyses can be used as a basis for refining the code. Although this approach takes a certain amount of time at first, it leads to significant reductions in code size after only a few iterations while still maintaining adequate execution speed. Of course, you cannot expect miracles here.

These analyses often show that Java compilers do not optimize as well as C or C++ compilers. This is primarily due to the fact that just-in-time (JIT) compilers are generally used in the PC world. Such compilers translate the Java bytecode into the machine language of the target system before it is run for the first time and optimize it in the process. Relatively low optimization of the source code makes it easier for a JIT compiler to generate target code quickly. Incidentally, this also clearly indicates that, despite many assertions to the contrary, the primary focus of Java is not the embedded-application environment, since JIT compilers cannot be used in this environment in the foreseeable future owing to limited processing power and memory space.

```
// Example of a somewhat suboptimal translation of Java into Java bytecode.
// The Java compiler uses the compute-intensive smul command for multiplication
// by 4 instead of using shift operations, which would be more efficient here.
// This optimization is especially beneficial in cases where multiplication
// is executed repeatedly in a loop or similar structure.

short x, y;               // x and y are variables of type short

x = (short) (y * 4);    // multiply y by 4
// Translation of Java source code into Java bytecode
// 00000100    16    sload #5      // load y
// 00000102     7    sconst_4      // load constant 4
// 00000103    45    smul          // multiply y by 4
// 00000104    29    sstore #4 x   // store in x

x = (short) (y << 2);  // multiply y by 4
// Translation of Java source code into Java bytecode
// 00000100    16    sload #5      // load y
// 00000102     5    sconst_2      // load constant 2
// 00000103    4D    sshl          // shift y left two places
// 00000104    29    sstore #4 x   // store in x
```

Most of the following suggestions are generally applicable to code-based smart card applications with or without interpreters. The specific characteristics of Java cards are featured prominently here owing to the widespread use of Java.

6.5.1 Data types

The Java Card specifications provide only a subset of the data types defined for normal Java. The available types are 'boolean' (true, false), 'byte' $(-2^7 \ldots 2^7 - 1)$, 'short' $(-2^{15} \ldots 2^{15} - 1)$, and optionally 'int' $(-2^{31} \ldots 2^{31} - 1)$. These data types are sufficient for implementation of commonly encountered specifications for smart card applications and do not cause any significant restrictions.

However, it is advisable to observe the following suggestions when using the individual data types. Depending on the specific case, this can occasionally lead to considerable improvements with regard to the program parameters of memory usage, execution speed and robustness.

Constants   Constants provide a simple means to ensure that certain items have fixed values in the entire program. Besides improving the readability and maintainability of the source code, this improves the quality of the code because constants cannot be overwritten unintentionally. However, the main benefit is that global constants declared using 'static final' do not take up scarce RAM space, but are instead stored in EEPROM as part of the program code.

If you use the modifier 'static' with a constant, you can also ensure that the constant will be stored only once in the applet or package, instead of duplicate constants being stored for each instance. Constants should be used extensively, since they yield better programs while providing a simple means to save RAM space.

```
// Various options for defining the instruction byte of a command

// Modifiable variable in nonvolatile memory;
// duplicated for each instance
public byte            INS_SELECT  = (byte) 0xA4;
// Nonmodifiable constant in nonvolatile memory;
// duplicated for each instance
public final byte      INS_SELECT  = (byte) 0xA4;
// Modifiable variable in nonvolatile memory;
// present only once for all instances
public static byte     INS_SELECT  = (byte) 0xA4;
// Nonmodifiable constant in nonvolatile memory (best solution);
// present only once for all instances
public final static byte INS_SELECT  = (byte) 0xA4;

// Typical instruction byte definitions for commands according to ISO/IEC 7816-4
final static byte INS_SELECT  = (byte) 0xA4;      // SELECT command
final static byte INS_VERIFY  = (byte) 0x20;      // VERIFY command

// Typical definitions of selected return codes
// PIN verification failed
final static short SW_PIN_FAILED =      (short) 0x63C0;
// Referenced data not found
final static short SW_DATA_NOT_FOUND = (short) 0x6A88;

// Typical definitions of selected data elements
// Length of smart card identification number
final static short LEN_SCID = (short)  4;
// DES key length in bytes
final static short LEN_KEY  = (short)  8;
// Pre-defined PIN; DEFAULT_PIN is a reference to an array
final static byte[] DEFAULT_PIN = {(byte) 0x12, (byte) 0x34, (byte) 0x56};
```

Initialization of variables   Global variables are automatically initialized to 0 by the Java runtime system when they are instanced. This causes numeric types to be initialized to 0, boolean types to be initialized to 'false', and references to be initialized to 'null'. If the default value matches the required initialization value, explicit initialization in the program code is not necessary. This saves a bit of memory space without incurring any additional risk.

Variables in RAM   Variables can be stored in RAM or EEPROM in a Java smart card. If they are located in RAM, their contents are not saved across sessions. By contrast, the contents of variables in EEPROM are saved across sessions. However, write accesses to RAM variables are on the order of 10 000 times faster than write accesses to variables in EEPROM.[1] In addition, you should bear in mind that the guaranteed lifetime of EEPROM as specified by semiconductor manufacturers is typically only 500 000 write accesses per page. Although the actual value can easily be 10 times higher under normal ambient conditions (temperature and supply voltage), if you rely on this you are are working outside the specification envelope, which in certain cases can quickly lead to premature failure of the smart card. Loop counts (to take one example) are definitely not suitable for storage in EEPROM as variables.

Global working variables   In well-structured software, specific local variables are used for each method. However, this requires a considerable amount of space on the stack and markedly increases processing time. In the case of Java Card, it is advisable to use global variables (valid for the entire applet) to the extent that is reasonably possible. These variables are defined and initialized once only in a central location, after which they can be used in the entire program, and they are stored in EEPROM.

Naturally, when you use global variables you must strictly ensure that they are not modified unintentionally by a method. Such modifications can be a major source of errors, which is furthermore difficult to track down. To the extent possible, you should avoid naming global working variables as 'work1, work2, work3'. More or less meaningful names such as 'len' or 'index' should be used instead. You can also explicitly identify them as global working variables (e.g. g_len or g_index). This at least gives you a minimum degree of protection against the risks associated with global variables.

```
public class Test extends Applet {
  ...
  // Specify global working variables

  // Reference to working memory for arbitrary data
  static byte[]      memory;
  // Number of entries in working memory
  static short       memory_NoOfEntries;
  // Package global working variable
  static byte        g_x;
  // Package global working variable for lengths
  static short       g_len;
  // Package global working variable for array indices
  static short       g_index;
  ...
```

Reusing variables   Reusing local variables is subject to the same considerations as with global variables. The theoretically correct approach would be to define a specific variable for each individual quantity. Owing to the amount of storage space this would require on the stack and the execution time required to generate and initialize the variables, you should use this approach sparingly. It is better to reuse available local variables for

---

[1] Access to a RAM cell typically takes two processor clock cycles. At a clock rate of 3.5 MHz without clock multiplication in the microcontroller, a byte can be written to RAM in approximately 600 ns. Writing to EEPROM is independent of the clock rate, and it usually takes around 3.5 ms without a prior erase operation. As Java Card objects are normally written using atomic transactions, at least two additional EEPROM write accesses are necessary. The ratio of the write times for the two types of access can be calculated roughly using these figures: $(2 \times 3.5\,\text{ms}) \div (600\,\text{ns}) = 11\,666$. This proves the rule of thumb that EEPROM accesses in a Java cards are around 10 000 times slower than RAM accesses

new purposes if they are no longer used in a particular location. It helps to assign each variable used for several purposes a name that clearly indicates how it is used in all cases.

Boolean   The logical data type 'boolean' is very useful in smart card applications, since these applications use a lot of yes/no decisions. However, this data type occupies 8 bits or 16 bits of memory, depending on the implementation, instead of only 1 bit. If you need to save space and require several boolean variables, you should consider using the 'byte' or 'short' data type instead. Up to eight true/false values can be stored in a 'byte' data type, or up to 16 in a 'short' data type. Bits can be set, reset or polled in variables of type 'byte' or 'short' using simple logical operations, so the necessary software overhead and additional risk of errors are not overly large.

```
// Storage-space optimized processing of true/false values in byte variables

// Declaration of constants
final static byte  BIT0 = (byte)  0x01; // 0x01 = 0000 0001
final static byte  BIT1 = (byte)  0x02; // 0x02 = 0000 0010
...

// Declaration of variables
byte bitfield;

// Set flag BIT0 in variable bitfeld
bitfield = (byte) (bitfield | BIT0);
// Further simplification: set flag BIT0 in variable bitfeld
bitfield |= BIT0;

// Clear flag BIT0 in variable bitfeld
bitfield = (byte) (bitfield & ~BIT0);
// Further simplification: clear flag BIT0 in variable bitfeld
bitfield &= ~BIT0;

// Check whether flag BIT0 in variable bitfeld is set
if ((bitfield & BIT0) != 0) {...}    // Bit0 is set

// Check whether flag BIT0 in variable bitfeld is cleared
if ((bitfield & BIT0) == 0) {...}    // Bit0 is not set
```

Arrays   Arrays are quite handy, and they can be used in all types of smart card applications. However, arrays are administered as objects in Java, which results in an additional storage overhead for administrative data and an additional time overhead for access, in part because a consistency check of the object, a firewall check, and a check of the array boundaries must be performed before each access. As a developer, you should always keep these Java VM overheads for arrays in mind.

Nevertheless, arrays are unavoidable in the specific application area of smart cards. There are two ways in which you can reduce access times for arrays. The first is to use the methods provided by the Java Card API for comparing arrays ('arrayCompare()'), copying arrays ('arrayCopy()' and 'arrayCopyNonAtomic()'), and filling arrays ('arrayFillNonAtomic()'). These methods run significantly faster than user-generated Java program code because they are implemented in the machine language of the target system, so you should always use them.

The biggest mistake you can make with regard to execution speed is to program a loop that copies data from one array to another byte by byte. In such a process, an array boundary check and firewall check will be performed before each array access.

The second way to reduce array access times is to use local cache variables to store one or more elements of an array when it is first accessed. All subsequent accesses can then use the cache variables instead of the variables in the array. The length of the array is typically stored in one of these variables, as accessing the array length instance variable takes significantly more processing time.

```
// Example of a time-intensive query of the array size instead of
// using the already known size data in the form of a constant
public class TEST extends Applet {
    ...
    // Array size for a universal data storage medium
    final static short SIZE_MEMORY = (short) (5);
    ...

    public static void install(byte[] buffer, short offset, byte length) {
        memory = new byte[SIZE_MEMORY];   // Generate a universal data storage medium
        ...
    }

    private void cmdSELECT(APDU apdu) {
        short x;
        ...
        // The length query can be avoided, since the value range is available via
        // the constant SIZE_MEMORY
        x = (short) memory.length;
```

Type conversions   Improper type conversion is by far the most common source of errors in programs for Java cards. This is primarily due to the fact that all numeric data types in Java are signed, while signs play a very modest role in currently used smart card specifications because only unsigned integers are normally found in these specifications. Interestingly enough, the reason behind the systematic use of signed numbers in Java is that both signed and unsigned variables are possible in the C language and this is a constant source of errors in C code.

Improper type conversions are often rather tricky to identify because there might be a limited region within which the conversion is entirely correct. For example, if a variable that can assume any value in the range of 0–255 is mapped into a variable of type byte, which has a range of −128 (1111 1111) to +127 (0111 1111), the conversion will be correct in the range between 0 and 127. However, the value of the 'byte' data type will be negative if the input value is greater than 127 owing to the fact that the most significant bit is set. As is well known, in Java the most significant bit represents the sign of a variable. An error of this sort may remain undiscovered in a program for a long time, but it can still cause enormous problems later on.

For this reason, the 'short' data type is often used in Java applets instead of the 'byte' type, since it can cover the range of 0–255 without any problem. Caution is nevertheless required when the 'short' data type is used to cover the positive integer range of two bytes, such as for indexing into an EF with a transparent structure, since the range of the signed data type short only extends to 32 767 ($2^{15} - 1$) instead of 65 535 ($2^{16} - 1$).[1] The data type 'int' can be used in such cases, but it is not supported by all Java cards. Alternatively, two variables of type short or a byte array can be used to construct a suitable variable type. The use of a byte array has the additional advantage that it can

---

[1] The maximum plausible size of a transparent file is 33 023 bytes (see Section 2.1.3). However, there are standards such as TS 51.011 that allow the file size to extend over the full range of values of a 2-byte index

easily be scaled to represent even very large integer values, depending on the size of the array. Although this may appear somewhat cumbersome, it is quite often used in practice.

Naturally, you should also check all arithmetic operations to see whether the type conversions performed quasi-automatically by the compiler correspond to the intention of the developer and do not give rise to unforeseen errors. The intermediate results of computations are particularly critical in this regard, and you should always examine them carefully for correct type conversion.

```
// Example of correct type conversion of a variable of type byte to a variable
//   of type short

// Store a received command APDU
byte[] cmd_apdu = apdu.getBuffer();

// Convert a 1-byte unsigned integer into a 2-byte signed short variable
// x: bit value of an integer, z: sign
// Content of cmd_apdu[ISO7816.OFFSET_P2]:                  xxxx xxxx
// Coding of cmd_apdu[ISO7816.OFFSET_P2]:                   zxxx xxxx
//                            ANDed value:  0000 0000 1111 1111
//              Result of the AND Operation: 0000 0000 xxxx xxxx
//          Assignment to the variable tag: zxxx xxxx xxxx xxxx

// Map the byte value of P2 in the command APDU to the short variable tag
short tag = (short) (cmd_apdu[ISO7816.OFFSET_P2] & (short) 0x00FF);

// Check whether the value of tag is lower than the allowable range
if (tag < (short) 0x0040) {ISOException.throwIt(ISO7816.SW_WRONG_P1P2);}

// Map the byte value of Lc in the command APDU to the short variable lc
short lc = (short) (cmd_apdu[ISO7816.OFFSET_LC] & (short) 0x00FF);
```

The following example can serve to illustrate the dangers of improper conversion. In a smart card application generated in Java, TLV-coded data was stored in a file. The coding of the TLV length parameter was implemented without any error, and access to the coded data in the file passed all blackbox tests without any errors in a smart card with 38 KB of memory. After a period of more than two years, it was decided to migrate the application to a smart card with 128 KB of memory. In principle, this is a simple task. The tests were updated accordingly, but much to everyone's surprise the smart card terminated processing with the indeterminate error message '6F00' ('Check error – no specified diagnosis') when more than a certain number of data objects were present.

A Java Card simulator was used to analyse the program in order to track down the reason for the unexpected behaviour of the smart card. However, it took a lot of intensive digging to discover the cause of terminated processing. It lay in the index to the data, which was implemented using a variable of type short. Using a smart card with more memory made it possible to store more data objects and longer data objects, with the result that an exception occurred if the file index was greater than 32 767. This was not possible with the card that was used originally because the memory was too small. What made this problem particularly tricky was that the error only occurred when the sum of the lengths of all the stored TLV data objects was greater than 32 767. After this was discovered, the data type of the file index was changed to int, which solved the problem. The pity was that two different applets were now necessary for the same application, depending on the memory size of the smart card, and one of them used the optional data type int.

Unused variables Variables introduced during the development process frequently become unnecessary later on after the program has been revised. For this reason, you should recheck the entire program code for unused variables after the software has been finished. Some development environments, such as Eclipse, now perform this step automatically. You can also check whether variables can be reused or existing global working variables can be used instead. Considerable amounts of memory space can sometimes be saved by making this check, which does not take all that long. Of course, all tests must be repeated in full after an optimization of this sort.

Integer data type The 'int' data type is optional according to the Java Card specifications, so it is not supported by some smart cards. This data type should not be used in applets that must be able to run in a wide variety of Java cards, since otherwise unpleasant surprises can crop up quite quickly.

6.5.2 Arithmetic operations

Since the early days of information technology era, a large number of tricks have been used to accelerate arithmetic operations. Probably the best-known trick is to use shift-left and shift-right operations to multiply or divide integer powers of 2. However, most computational optimizations of this sort have long since been incorporated into modern compilers, so developers no longer have to use them explicitly.

The biggest danger in accelerating computations in this manner is that it can cause the source code to be too close to the assembly-language level under certain conditions. This harbours the disadvantages of reduced comprehensibility and an increased likelihood of errors. The best way to deal with this is to have experts occasionally review the generated Java byte code at critical locations during the development process. If it turns out that the compiler does not produce an optimal translation of the Java source code for arithmetic operations, the code can always be optimized manually as necessary. Naturally, this presumes that a suitable method is not already provided by the Java Card API. The following program code for querying an integer to determine whether it is even or odd provides a simple example of a possible optimization.

```
// Two different implementations of functionally equivalent modulo calculations,
// which result in different timing characteristics. The srem bytecode takes
// slightly more processing time in the Java Card VM than the sand bytecode.
// This optimization is especially beneficial in cases where the calculation is
// executed frequently in a loop.

// Declaration of variables
short x;

if (x%2 == 0 )        {...}        // x is an even number
>>00000100   16    sload #4 x
>>00000102    5    sconst_2
>>00000103   49    srem           // calculate the remainder using short
>>00000104   61    ifne 108

if ((x&0x01) == 0) {...}          // x is an even number
>>00000100   1F    sload_3 x
>>00000101    4    sconst_1
>>00000102   53    sand           // logical AND using short
>>00000103   61    ifne 107
```

Extensive calculations in a single line of source code take up somewhat less memory space than the same calculations spread out over several lines. However, the difference is rather small, so this form of optimization is only worthwhile in borderline cases because it sometimes makes debugging more difficult.

```
// Calculating a value using two different methods
// with different memory usage

// Declaration of variables
short index, startindex, offset, lenofindex;

// The following calculation requires 13 bytes of program memory
index = (short) (startindex + offset);
index = (short) (index + lenofindex);
//00000100   16    sload #7 startindex
//00000102   16    sload #8 offset
//00000104   41    sadd
//00000105   29    sstore #6 index
//00000107   16    sload #6 index
//00000109   16    sload #9 lenofindex
//00000111   41    sadd
//00000112   29    sstore #6 index

// The following calculation requires 9 bytes of program memory
index = (short) (startindex + offset + lenofindex);
//00000100   16    sload #7 startindex
//00000102   16    sload #8 offset
//00000104   41    sadd
//00000105   16    sload #9 lenofindex
//00000107   41    sadd
//00000108   29    sstore #6 index
```

6.5.3   Control structures

If you analyse typical Java Card applications, you will see quite clearly that 'if–then–else' and 'switch' statements are the most commonly used control structures. It is thus worthwhile to give them some attention, since well-considered use of these control structures can yield benefits in terms of execution speed and memory usage.

If–then–else statement   The 'if–then–else' statement is the most commonly used control structure. It thus claims a corresponding proportion of the total memory space occupied by the application, so it pays to use an optimization measure here to save space.

In many cases, the 'else' branch of the statement can be omitted by assigning a suitable initial value to the variable concerned. This consists of assigning the normal value to the variable before executing the 'if' instruction and then modifying it in the 'if' branch only if necessary. This makes the 'else' branch of the conditional statement redundant. However, this approach must be used judiciously, since otherwise the readability of the code will suffer.

```
// Two different implementations of functionally equivalent
// conditional statements and their associated memory requirements

// Declaration of variables
short index, buffer;

// The following computation requires 12 bytes of program memory
```

```
if (index == MAX_VAL) {
  buffer = FULL;
} // if
else {buffer = EMPTY;}
//00000100    16    sload #10 index
//00000102     8    sconst_5
//00000103    6B    if_scmpne 110
//00000105     4    sconst_1
//00000106    29    sstore #11 buffer
//00000108    70    goto 113
//00000110     3    sconst_0
//00000111    29    sstore #11 buffer

// The following calculation requires 10 bytes of program memory
buffer = EMPTY;
if (index == MAX_VAL) {
  buffer = FULL;
} // if
//00000100     3    sconst_0
//00000101    29    sstore #11 buffer
//00000103    16    sload #10 index
//00000105     8    sconst_5
//00000106    6B    if_scmpne 111
//00000108     4    sconst_1
//00000109    29    sstore #11 buffer
```

Switch statement   Switch statements execute faster than nested 'if–then–else' statements, and they also require less memory space. This is because Java provides specific bytecodes for this form of branching ('slookupswitch' and 'ilookupswitch'), and they execute faster than equivalent 'if–then–else' conditional statements. Another advantage is good readability even when a large number of conditions are tested. Consequently, 'switch' statements are the preferred solution for testing multiple conditions.

```
// Core routine of a simple command interpreter of a Java Card application
// constructed using 'if-then-else' statements. These conditional statements
// require more memory than a functionally equivalent 'switch' statement,
// and they are slower due to the repeated array queries.
if (cmd_apdu[ISO7816.OFFSET_INS] == INS_SELECT)
  // SELECT command; call SELECT command processing
  cmdSELECT(apdu);
else if (cmd_apdu[ISO7816.OFFSET_INS] == INS_VERIFY)
  // VERIFY command; call VERIFY command processing
  cmdVERIFY(apdu);
else if (cmd_apdu[ISO7816.OFFSET_INS] == INS_PUTDATA)
  // PUT DATA command; call PUT DATA command processing
  cmdPUTDATA(apdu);
else if (cmd_apdu[ISO7816.OFFSET_INS] == INS_GETDATA)
  // GET DATA command; call GET DATA command processing
  cmdGETDATA(apdu);
else
  // Non-supported command
  ISOException.throwIt(ISO7816.SW_INS_NOT_SUPPORTED);

// Core routine of a simple command interpreter of a Java Card application
// constructed using a 'switch' statement
switch(cmd_apdu[ISO7816.OFFSET_INS]) {
  case INS_SELECT:      // SELECT command
    cmdSELECT(apdu);    // Call SELECT command processing
    break;
  case INS_VERIFY:      // VERFIFY command
    cmdVERIFY(apdu);    // Call VERIFY command processing
    break;
  case INS_PUTDATA:     // PUT DATA command
    cmdPUTDATA(apdu);   // Call PUT DATA command processing
    break;
  case INS_GETDATA:     // GET DATA command
```

```
      cmdGETDATA(apdu);    // Call GET DATA command processing
      break;
   default :               // Non-supported command
     ISOException.throwIt(ISO7816.SW_INS_NOT_SUPPORTED);
 }  // switch
```

Exceptions   The exception handling capability of Java is a powerful tool that can enhance program stability if it is used carefully. It also allows the code of the actual function
to be separated from the error handling code, which is highly desirable. However, this
benefit comes at the price of a larger program size, since all information regarding exception handling must of course be made available to the Java VM. In addition, exception
handling is significantly slower than equivalent control structures, in part because it requires at least two firewall checks. Injudicious use of try–catch statements for exception
handling can also degrade the readability of the source code. In summary, in our opinion
the exception handling mechanism should only be used for truly critical special cases,
and it should never be used as a control structure for normal processes.

6.5.4   Methods

Structuring methods properly is a critical factor for generating robust, error-free programs
for smart cards. The most important aspect is that all methods must be structured defensively. They must be stable and able to defend themselves against invalid or incorrect
input data as well as possible. All significant exceptions inside methods must be handled
appropriately. Ideally, methods should also be structured so that an orderly fallback is
possible in non-critical cases without terminating the transaction. If an intolerable error
nevertheless occurs, the method must return a meaningful error message without thereby
creating an opportunity for an attack. Critical variables should be checked at suitable
locations inside methods (such as after complex calculations) to determine whether they
are within their correct ranges. In addition, reliable measures must be taken to ensure
that incorrect return values can never occur under any conditions.

Calling hierarchy   If they are properly structured, deeply nested calling hierarchies
can reduce development costs and produce source code that is easier to understand.
However, they require additional program memory, processing time, and stack space.
You should thus critically consider, on a case-by-case basis, whether these disadvantages
are outweighed by the benefits – or whether it would be better to use a flat calling
hierarchy.

Method attributes   If a method does not have the attribute 'static', at run time the Java
VM must determine dynamically whether the method is to be invoked as an overwritten
derived class or as the base class. This takes a corresponding amount of processing
time and retards program execution. Consequently, methods should be declared with the
additional attribute 'final' as much as possible. Of course, this means they can no longer
be overwritten, but this generally does not create any restrictions for applications that
run in the smart card environment. If overwriting becomes necessary at some later point
in time, you can always revise the corresponding method declaration.

Return points   A standard rule of modern software development says that each method
must have only one return point instead of several return points. The objective of this

is to improve the clarity and comprehensibility of the source code. However, in some cases this results in additional control structures or auxiliary variables, which is not a disadvantage in PC programs.

In contrast to PCs, the amount of memory available in a Java card is quite limited, so non-essential conditional statements and variables should be avoided. Consequently, as a professional developer of smart card software, you should regard the rule 'only one return point per method' as a friendly suggestion for producing flawlessly structured code instead of as a hard and fast rule.

Recursion   There are several algorithms (such as Quicksort) that can best be implemented using recursion. However, this technique should not be used in software for smart cards, since any problem in the recursion path will cause all available memory to be used. Under unfavourable conditions, this can render the card useless.

A second important argument against recursion is that the amount of memory available in smart cards is anything but generous, so it can easily be exhausted with only a few recursion loops. Recursion can be replaced by other methods that are much more suitable to the requirements imposed on smart card software.

There is also a third reason why recursion should not be used. Unfortunately enough, the Java Card specifications do not mention anything about the minimum size of the stack, with the result that a wide variety of stack depths are found in practice in various types of Java cards. Liberal use of the stack should thus be subjected to careful examination, since otherwise porting the software can be unnecessarily difficult or even impossible in unfavourable circumstances.

6.5.5   Applets

Programs are always loaded into Java cards in the form of packages. A package can contain one applet, several applets or no applets. If it contains applets, they are assigned to a group context. As a result, all instances of the classes of a packet are located in the same context. As is well known, dynamic downloading of classes is not possible with Java cards. An advantage of having all the applets of a package in the same context is that no firewall checks are made for accesses inside the group context, so these accesses are significantly faster than accesses outside the package boundary. Incidentally, the number of classes in Java Card applets is almost always in the single-digit range.

Object orientation   Along with C++, Java is one of the best-known object-oriented programming languages, and it implements all the features of the object-oriented model. All object-oriented methods are also possible in principle with Java Card. However, you should bear in mind that the processing power and memory capacity of a smart card is nowhere near that of a PC, and this situation will certainly not change in the foreseeable future.

For this reason, most object-oriented methods should be used with caution in smart cards, since they can very quickly cause the amount of available memory to be exceeded or make program execution much too slow. Object-oriented approaches such as inheritance and superimposition of methods, polymorphism, abstract classes, and data encapsulation should be used only in small measure or not at all. Object-oriented design patterns such as those described by Gamma et al.,[1] which are otherwise very useful, can also be used only sparingly in smart cards. Ideally, you should try to use a style based on procedural programming languages such as Basic and C. This also corresponds to the underlying philosophy of this entire section.

The 'new' operator   The 'new' operator can be used to generate an object from a declared class. The storage space for the object header and object body is taken from the heap. This means that each new object reduces the size of the free heap according to the memory requirements of the object. If the generated object is no longer needed later on, this is recognized by the garbage collector in commonly used implementations of Java on PCs, and it releases the memory allocated to the object so that it is again available on the heap. Similar considerations apply to transient memory (i.e. RAM) that is no longer referenced.

Unfortunately, a garbage collector is specified as optional in Java Card, so in certain cases it may not be possible to subsequently release the allocated memory. Although many Java Card implementations do provide a garbage collector, as an application developer you cannot assume that it will be present in every Java smart card. Incidentally, the garbage collector must be invoked explicitly in the applet by using the 'requestObjectDeletion()' method, and it only become active when the 'process()' method has been completed, which means when smart card command has been fully processed.

Consequently, the 'new' operator must be used very judiciously in order to avoid using up all available heap memory, which would ultimately cause the smart card to persistently return an 'insufficient memory' error message. It is thus a good idea to generate all new objects in a single, central location in the program code. The best place for this is the install method or constructor method of an applet. You can reliably assume that these methods will be invoked only once when the applet is installed. This also ensures that all necessary objects will be generated only once. In addition, grouping all object generation actions in one place makes it easy to check all new operators to ensure that objects are not generated in other places as well.

```
// Sample structure of the install method of a Java Card applet with the name
// TANGen. All objects required for the entire applet are generated centrally
// in this method.

public static void install(byte[] buffer, short offset, byte length) {
    // Generate a byte array with length LEN_SCID in nonvolatile memory
    scid = new byte[LEN_SCID];

    // Generate a byte array with length LEN_SEED in nonvolatile memory
    seed = new byte[LEN_SEED];

    // Generate a byte array with length SIZE_MEMORY in nonvolatile memory
    memory = new byte[SIZE_MEMORY];

    // Generate a transient byte array with length LEN_ARRAY in volatile
    // memory. This array is deleted after the applet is deselected.
    workarray = JCSystem.
            makeTransientByteArray(LEN_ARRAY, JCSystem.CLEAR_ON_DESELECT);

    // Register the applet in the Java Card Runtime Environment (JCRE)
    new TANGen().register();
}   // Install
```

---

[1] See Gamma *et al.* (1994)

In the above programming example, all objects are created before the 'register()' method is invoked, which is also correct in principle. However, here you should bear in mind that the commit buffer in the Java card may not be large enough to create very large objects, which will cause the process to abort. Consequently, in such cases you should first use the 'getMaxCommitCapacity()' method to check the size of the available commit buffer or delay object creation until after the 'register()' method has been invoked.

Java Card API   Java cards have an API (application programming interface) that is optimized for the needs of smart card applications.[1] The obligatory portions of this interface must be supported by all Java cards, and their functionality is optimized for the smart card environment. They provide a Java interface to the calling program, and the core routines behind this interface are executed directly by the processor without any involvement of the Java VM. This makes relatively high execution speeds possible.

For this reason, it is important for developers to be familiar with the entire functionality of the Java Card API and utilize it in their own applications. Functionally identical user-generated code in Java should be avoided because it will be slower and consume more memory space than the methods provided by the API. On the other hand, before you use a proprietary interface that is only available in a specific smart card operating system, you should consider that this will complicate subsequent porting of the Java applet. Decisions regarding utilization of such interfaces should be made on a case-by-case basis after carefully weighing all the potential advantages and disadvantages.

Using API methods not only saves memory space and increases execution speed, but also improves security against attacks because the security-critical processes run in program code that is protected by the API. The best example of this is using the 'check' method to compare a transferred PIN with a stored PIN. This method takes into account typical forms of attack on the PIN based on analysing timing or current consumption characteristics, while at the same time protecting the PIN error counter against manipulation.

If the Java card provides an interface for accessing and managing an ISO/IEC 7816-4 file system, this interface should be used. All Java SIM and USIM cards provide an interface of this type, which is standardized in TS 101 476. The advantages in terms of code size and speed relative to a user-generated file system coded in Java are enormous.

Transactions   The Java Card Runtime Environment (JCRE) provides several methods for atomic transactions. They can be used without any additional programming effort to ensure that certain data is always written to nonvolatile memory in its entirety or else not at all. This is important in some scenarios for files and data objects, since it reliably excludes data corruption even in the event of an unforeseen power interruption. However, these atomic transactions are significantly slower than non-atomic transactions, and they reduce the useful life of the commit buffer. The transaction depth, which means the size of the commit buffer, is unfortunately not stated in the Java Card specifications. As a result, different types of Java cards have different buffer sizes, which can lead to portability problems.

Consequently, you should always keep an eye on the size of the commit buffer when using atomic transactions. One easy way to do this is to use the 'getMaxCommitCa-pacity()' method, and all possible exceptions related to the commit buffer should also

---

[1] See JCAPI (2003)

be handled appropriately. Incidentally, it is possible to insert non-atomic transactions between atomic transactions defined using the 'beginTransaction()' and 'commitTrans-action()' method calls. This represents a reasonable compromise in some cases. However, you should always give preference to non-atomic transactions to the extent permitted by the smart card application concerned.

Class hierarchy   Derived classes are a standard feature of object-oriented software development, but they impose distinct demands on the two critical aspects of memory space and processing power. For this reason, you should have a good general idea of how a derived class is created in Java. Besides the administrative data, the compiler generates a method table (among other things) for the new class. At runtime, the Java VM must search this new table when a method is accessed. If the method of the derived class is not listed in the table, the method table of the next higher-level class is searched. The search proceeds through the class hierarchy until the method is finally found. This search process takes a significant amount of time, which reduces the execution speed of the routine accordingly.

In summary, it can be remarked that you should try to achieve a class hierarchy that is as simple and flat as possible, and you should use derived classes only in exceptional cases.

Redundant program code   The considerations regarding unused variables that creep into the code during the development process are equally applicable to redundant and dead program code. The software development process is rarely perfect, so redundant code is often present in programs. Owing to the severe memory space limitations in Java Card applications, it is advisable to refactor the code at the end of the implementation phase to combine identical or similar program codes. This process must be performed with considerable care.

Chapter 7

Operation Patterns

Several important considerations must be borne in mind when deploying and operating a smart card system, as otherwise considerable difficulties can arise. This begins with the personalization phase when data is entered into the smart cards before they are issued. In the operational phase, suitable migration concepts are necessary for all systems in which cards can only be replaced incrementally, and the system must be monitored for signs of instability or attacks. The wealth of options provided by the online and offline behaviour of smart cards can be put to good use here, but they also require certain measures to be taken to achieve stable system operation.

7.1 Initialization and Personalization

After development of the smart card application has been finished and all testing and release activities have been completed, the next major step is producing the smart cards. This includes producing the printed smart card bodies and embedding the chips. This step is usually handled by a card manufacturer. More than 100 billion cards are produced each year by card manufacturers.

The next production step is card-specific printing of labels and images, as well as loading data and programs into the nonvolatile memory of the microcontroller. This is also the job of the card manufacturer when large numbers of cards are involved, but it can also be done manually or semi-automatically with a small-to-medium number of cards. The number of cards for which this is suitable can easily extend up to several tens of thousands of cards per year. Desktop machines that employ the thermal transfer process are typically used for card-specific printing because of their operational flexibility.

The decisive step from an informatics perspective is electrical initialization and personalization. All generic data – which means all data that is the same for all cards – is written to memory during initialization. All data related to individual cards or users is written during personalization, which is why this step is also called *individualization*. This distinction is especially significant when large numbers of cards are involved, since the generic data can be written during the first step without accessing internal or external databases, while loading card-specific data into the cards can be postponed to the personalization step. In this way, a distinct informatics boundary can be drawn between data that is common to all cards (generic data) and data that is specific to individual cards.

Initialization and personalization are normally performed in a secure area. This has the enormous advantage that communication with the smart cards does not have to be secure, which is considerably simpler and somewhat faster than data transmission using Secure Messaging.

There is one more test that must be made before the data is loaded. Although it is a trivial test, it is nevertheless very important. It consists of checking that the right smart cards are being initialized or personalized. The object here is to ensure that the intended smart card microcontroller and the right version of the smart card operating system are being used. This is highly important, since there might not be enough memory available in the subsequent operational phase if the wrong microcontroller is used, and some versions of the operating system may not provide all the functions required by the application.

Table 7.1 summarizes the sequence of activities involved in card initialization and personalization.

Table 7.1   Basic steps for initializing and personalizing a smart card

Step	Activity
1.	Check that right microcontroller and operating system are present in the card (e.g. GET DATA)
2.	Check the authenticity of the smart card (e.g. MUTUAL AUTHENTICATE)
3.	Initialization (program loading: LOAD and INSTALL, file creation: CREATE, loading constants: UPDATE BINARY, UPDATE RECORD and PUT DATA)
4.	Personalization (UPDATE BINARY, UPDATE RECORD and PUT DATA)

Here it is important to ensure that the smart card is genuine before loading data into it. This can be done by unilateral or mutual authentication. If this check were omitted, an attacker could slip a manipulated card into the process and have it loaded with real data. Later on, the attacker could read this data from the fake card and produce clones of the card. For this attack to succeed, the attacker must have access to the personalization process and must be able to introduce a manipulated card into the process and retrieve it without being detected. This is rather difficult to achieve. The procedure described above is an easy way to make this form of attack considerably more difficult, and it is common practice in all initialization and personalization processes.

Incidentally, verification of the authenticity of smart cards relative to the outside world requires more than just passing a secret password to the smart card in order to check whether the card is genuine. This mistaken approach crops up repeatedly, which is why it is mentioned specifically here. A secret password is ineffective in this context because preparing a manipulated smart card that acknowledges every password it receives as correct is trivially easy. Consequently, authentication using a challenge–response process is essential here.

Initialization is a simple process from an informatics perspective, since all the smart cards can be processed the same way. During initialization, standard commands (LOAD, INSTALL and CREATE) are used to load the software into the smart cards and create the necessary files. If common (generic) data must be written to the cards, this is done during the initialization process using standard commands (UPDATE BINARY, UPDATE RECORD and PUT DATA).

During personalization, the same commands are used to load individual data into each of the smart cards. Response data is generated during the personalization process for the system operator, which requires this data in order to integrate the newly produced smart cards into the system and put them into service.

There are two typical scenarios with regard to the structure of the personalization process. With a pure offline structure, the personalization computer reads the personalization data from a database and loads it into the smart card by using suitable commands such as UPDATE BINARY, UPDATE RECORD and PUT DATA. This process is logged individually for each card, and the logged data is stored in another database together with the response data for the system operator. This is shown schematically in Figure 7.1.

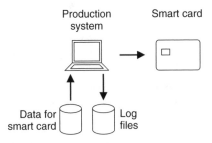

Figure 7.1 An arrangement for offline personalization of smart cards. This arrangement is commonly used in the industry

However, an online scenario is also used in certain cases. In this scenario, the personalization computer is linked to an external system. In the simplest version, the external system receives the response data from the personalization process, so the smart cards are known to the system after personalization and can be used immediately. This scenario is typically used for in-shop personalization where customers receive their personalized cards immediately after the process is completed.

In another version, personalization data is obtained from a higher-level system in real time as needed. Here again the typical use case is a shop where cards are personalized immediately before being sold, although in this case the personalization data is obtained from a central system as illustrated in Figure 7.2.

Figure 7.3 shows yet another form of online personalization. In this version, data held locally in the personalization system is sent to another entity that processes the data and then returns the result to the actual personalization process. This version is commonly used for signature cards with public keys that are signed by a trust centre before being loaded into the smart cards.

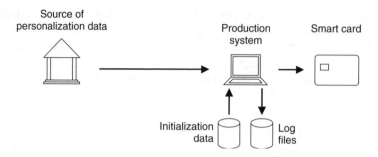

Figure 7.2   An arrangement for online personalization of smart cards, in which the personalization data is obtained from a higher-level system

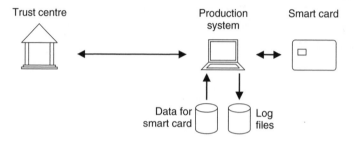

Figure 7.3   An arrangement for online personalization of smart cards in which the personalization data is supplemented by data provided by a higher-level system (such as a trust centre)

Which of these versions is used depends on the system architecture and specific operational parameters. A pure offline solution is easy to install, but it does not give the background system any direct means to monitor the number of cards that have been produced. Online solutions require a reliable, secure link between the background system and the personalization system, but they repay this additional expense with a direct stream of information about every personalized card.

Another aspect of personalization that must be borne in mind is that it must be possible to check all data written to the smart cards after personalization. These production checks must not be able to modify the data in the smart cards. In stable processes, these production checks are only performed on a sample basis, but they can also consist of full testing of every card if the quality of the personalization process is inadequate. The data dictionary[1] provides a suitable basis for this, since it contains a list of all data elements loaded into the smart cards.

This testing can be difficult in practice in some cases, since smart cards in particular contain a large amount of data that cannot be read afterwards or is not allowed to be read afterwards, such as passwords and keys. To enable testing to be performed despite this restriction, indirect forms of testing can be used. For instance, authentication can be used to check keys and VERIFY can be used to check PIN codes. As the personalization

---

[1] See Section 5.1

system possesses all the secrets of the smart cards concerned, these tests can be performed without any difficulty.

However, you must bear in mind that if there is a counter for the authentication key, it may not be set to its normal initial value when the cards are issued if this sort of indirect testing is used. If random tests are performed, the state of the authentication counter may not be the same in all of the issued smart cards, although the counters will usually differ by only 1 or 2. In some cases, this must be taken into account in the background system.

As every test of personalized data takes a certain amount of time, here again only critical or essential data should be checked. Tests performed on completion of personalization do not check the functionality of the smart card operating system, but only whether all the data was written correctly to the memory of the smart card. Checking the operating system at this point would take much too long.

As a supplementary security measure, the cards can be checked for the absence of test commands as a final step after the previously described test. This is a final security measure to intercept any overlooked test commands or test data before the cards are issued. In principle, this security measure is redundant, since test commands and test data should have already been weeded out in the previous processing steps. Nevertheless, it can be used here as a final check.

7.2 Migration

When a new smart card system is being deployed, emission of the smart cards usually begins several weeks or few months before the system is actually put into service. In the case of a smart card system with several million cards, for example, card emission typically takes three months. At the end of this interval, all the smart cards have been issued to the users and they can be used with the terminals after a defined starting day. This is a typical startup situation, and it is relatively unproblematic because all the elements in the field are present in well-defined versions and it is relatively easy to keep track of the mutually compatible components.

The situation becomes more difficult when a second generation of smart cards is issued. The cards of the second generation usually incorporate functional modifications and extensions with various degrees of significance. In addition, the new generation of cards may not be fully compatible with the previous generation. Nevertheless, the new cards must be usable immediately, since users will want to use them as soon as they receive them. This leads to a short interval during which two card generations – the old one and the new one – are in operational use, as illustrated in Figure 7.4. This means that the terminals and the background system must be able to handle both generations of cards for a certain length of time, which has a considerable impact on both of these components.

The interval during which both generations must be supported can easily extend over several years, depending on the individual system. This arises from the need to provide replacements for stolen, lost and defective cards, since replacement cards must be drawn from the current card generation until the new generation is available. As replacement

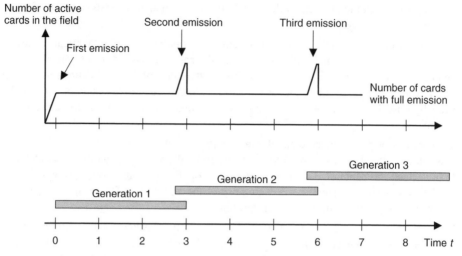

Figure 7.4   Migration timeline for successive generations of smart cards. The upper chart shows
the number of usable cards in the field versus time. The lower chart shows the
individual card generations and the overlaps during generation changes

cards are normally issued for the full normal term of validity for cost reasons, this means
that some of them can be used for nearly the entire term of validity of the subsequent
generation.

An important consideration here is that as soon as a generation change has started, all
replacement cards must be drawn exclusively from the new generation. Otherwise you
will suddenly find yourself in a situation where the terminals and background system
must support three different generations of cards when the time comes for the next
generation change.

This can be illustrated by a numerical example with a typical large smart card system.
Assume that a generation change occurs every three years and an emission phase lasting
three months occurs at the end of each generation. If a card that is lost one month before
the end of its term of validity is replaced by a card belonging to the first generation,
the replacement card will naturally be valid for three years after its date of issue. This
means that the replacement card will still be usable during the next generation change,

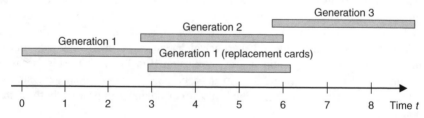

Figure 7.5   Migration from one smart card generation to a new generation with replacement
cards being issued during the emission phase of the next generation. This should be
avoided, as otherwise three different generations of smart cards will be in use during
the emission phase of the subsequent card generation

since the system was in the middle of a generation change when the replacement card was issued. As a result, three generations of usable smart cards will be present in the field during the next generation change. Figure 7.5 on the facing page illustrates this in graphic form.

If you wish to avoid this situation, you must always ensure that suitable measures are taken during migration from one card generation to the next. Naturally, the above considerations are equally applicable to other components of a smart card system, such as the terminals, the background system, and the entire key management system.

7.3 Monitoring

The background system of a smart card application must look after many different tasks. One of them is monitoring system operation. This primarily involves keeping an eye on two aspects: system integrity and attacks.

7.3.1 System integrity

Although the integrity of a smart card system can be upset by attacks, interface problems and faults in the system components are more common causes of integrity deficiencies. Nevertheless, a puzzling problem in the system integrity domain that seems to defy analysis can easily be a sign of an attack.

Integrity checking involves constant monitoring and statistical evaluation of essential system parameters. If the measured values lie below defined warning limits, no action is necessary. If a warning limit is exceeded, the corresponding parameter should be put under continual observation, but no immediate action is necessary. If an intervention limit is exceeded, the system administration organization must take immediate action. The necessary actions depend on the parameter being monitored, and naturally on the specific application.

This can be illustrated nicely by taking electronic purse systems as an example. If the number of cancellations lies below the warning limit, it is obviously not necessary to take any action. If the number of cancellations rises above the warning limit, it is advisable to monitor the transactions in question more closely. If the intervention limit is exceeded, a detailed analysis of the cause must be undertaken immediately. In non-critical situations, the limit may be exceeded because of 'natural' causes, but other causes are also conceivable – such as a particular type of terminal with a user interface that is so poorly designed that it regularly provokes cancelled transactions. This sort of situation should be detected by the background system, after which suitable corrective measures can be taken. However, it is also possible that an attacker has been experimenting with a terminal in an attempt to manipulate transactions.

This example also shows that the defined warning and intervention limits are strongly dependent on the monitored parameters and other system characteristics. As a rule, these limits are determined empirically during the trial operation phase and adjusted as necessary according to circumstances during the entire operational life of the system.

Another illustrative scenario can be taken from actual practice in payment systems. Two important parameters in such systems are the total balance of the system and the balances of the individual cards. The total balance must never go below zero, since this would mean that the amount of money that has been spent is greater than the amount previously loaded into the cards. In the individual cards, the sum of the load amounts must always be greater than or equal to the sum of the payment amounts. If this is not the case, the card in question can be loaded independently of the system. This means someone has figured out how to create electronic money.

Although these checks and this sort of monitoring sound rather simple in theory, they are tricky in practice. This is because large systems rarely work entirely online, so they usually do not have current transaction data constantly available for analysis in the background system. Many systems work with a large number of terminals that report their data to the background system only once a day or even only once a week. This time delay makes it difficult to reliably check system integrity continuously. Under these conditions, it may not be possible to detect problems in the smart card system until quite a while after they arise.

7.3.2 Attack detection

Massive attacks on the entire system can usually be detected by system integrity monitoring, although this is not always possible if the attacker takes a sufficiently clever approach. Attack detection must ensure that typical forms of attack on the smart card system are identified. With regard to the smart cards, it must be possible to detect clones and cards created outside the system. Clones are duplicates of genuine cards, and they exhibit all the data and characteristics of the original cards. Clones can be detected readily by analysing when and where cards are used and the patterns of use of individual cards in order to reveal irregularities.

Analyses of this sort are widely used in systems that use magnetic-stripe credit cards. For instance, it is fairly common practice to check whether a particular card has been used at the same time in different locations or whether it is physically possible to travel from one payment location for a particular card to another payment location for the same card within the time interval between successive payments. For example, if a credit card is used to purchase a pair of leather shorts in Munich at 12:00, a Tyrolean hat in Salzburg at 12:30, and a red-checked shirt in Tokyo at 13:00, this certainly isn't a case of a jet-setting visitor to the Munich Oktoberfest, but instead a couple of swindlers with cloned cards. In this case, the data of the original card has been transferred electronically to the locations where the clones in question were produced and used to make payments. However, it is difficult to use the information available to the background system to determine which one of the cards is the original. The usual defence measure for this sort of attack is to immediately block the card and enquire with the registered card holder.

Another method is to analyse patterns of use. If the genuine cardholder of a credit card has always used the card to make purchases of around a certain amount and the card is suddenly used to purchase an item costing several times the usual amount, the card may have been stolen or cloned. Here you can start to see how difficult attack detection can be in practice. It is naturally possible that the purchase was made by the genuine cardholder instead of by a crook. It is usually difficult to judge whether fraud is involved when deviations from an established pattern of use occur. Furthermore, the cardholder will be quite annoyed if the card is blocked without justification, and he or she may even cancel the card agreement. Naturally, the technique of analysing patterns of use can also be used for other smart card applications.

New cards produced by attackers, which means cards that have never been officially issued but are nevertheless regarded as genuine by the system, pose a significantly higher threat to system security than clones. With a small system that operates online, it is possible to use a whitelist and check every card that is used in the system against the list. This is easy to do because all the terminals operate online. If a card presented to the system is not on the list, it is not known to the system and the transaction is terminated.

However, this method is difficult to use in systems that employ offline terminals, since the white lists in the terminals must be updated regularly. Aside from the amount of data this involves, it is often impossible to maintain such lists current because the terminals are not always accessible online.

The most commonly used defence measure is to maintain a copy of the most important smart card data in the background system. In the case of an electronic purse system, this is the card balance of each card, and the copy of the balance of each card is called a *shadow account*. This information can be used to check submitted transactions for plausibility. In addition, a counter can be incorporated in each smart card to force it to conduct an online transaction when the limit count is reached. However, this approach can create problems with user acceptance if there are terminals in the system that cannot establish an online link with the background system due to their design. After all, card users want to be able to use their cards without being bothered with the details of system operation.

In summary, it can be concluded that a very good level of attack detection can be achieved in systems that operate online. In systems that work with offline terminals, which may also experience long delays between data acquisition and data transmission, there are significant difficulties associated with attack detection. Finally, it is appropriate to remark here that attackers do not always limit their efforts to manipulating smart cards, but can also attempt to manipulate terminals and the links between the terminals and the background system. The smart card system must be able to recognize these forms of attack with an acceptable probability of successful detection.

Chapter 8

Practical Aspects of Smart Cards

This chapter provides an overview of typical non-technical aspects of the practical use of smart card systems. The first aspect is user acceptance of smart cards, which has already confronted several highly vaunted systems with the hard facts of real life in a rather blunt manner.

Another aspect is proper high-level design of the smart card system. With some potential configurations, it is almost a foregone conclusion that the resulting system will be difficult to implement or operate. By contrast, intelligently selected mechanisms can considerably improve the chances of success of a smart card system. Similar considerations apply to dealing with problems that occur during the operational phase of a smart card system. If the persons involved act in a sufficiently professional manner, even relatively difficult situations can be mastered without major loss or damage.

Although these aspects are largely nontechnical in nature, they most certainly deserve the attention of developers and engineers, since even the most sophisticated smart card system is not good for much if it is not accepted by its intended users. It is also advisable to avoid focusing your thoughts and actions exclusively on the smart cards, since the other components of the overall system are also quite important. To give your smart card system a chance of ultimately succeeding, you must also give at least a certain amount of attention to 'soft' factors, such as human customs and behaviour.

8.1 Acceptance

The primary reason for the failure of many smart card systems has nothing to do with technical problems, but instead with the fact that they do not provide sufficient benefits for the involved parties. As the issuer of the smart cards, the system operator must bear the majority of the investment, which is why the system operator devotes considerable attention to ensuring that the system is at least worthwhile from its perspective. The situation is more difficult with the other parties, which in many cases can derive few, if any, benefits from the system. The primary issue that must be considered here is that the main interest of the potential users of a smart card system is the perceived direct

Smart Card Applications: Design Models for using and programming smart cards   W. Rankl
© 2007 John Wiley & Sons, Ltd

benefit, rather than a multi-page abstract treatise on the subject of 'What the new smart card system means for you'.

This can be illustrated quite dramatically with a few examples. In many West European countries, the legal framework necessary for using signature cards and digital signatures in parallel with conventional handwritten signatures was created in the 1990s. Many official organizations and system operators foresaw considerable direct benefits from such an arrangement, since it would facilitate automating a large number of business transactions and the signatures would be legally binding. Unfortunately, nobody gave any consideration to the fact that the potential users of signature cards did not perceive any significant personal benefit from using them. In the worst case, users were even expected to pay for the cost of acquiring signature cards and operating the associated system. In addition, there are many situations in which the envisaged card users have absolutely no interest in legally binding signatures and prefer well-established practices in the legal grey zone.

In all the euphoria about signature cards, people also overlooked another aspect: the necessary infrastructure is technically complex, and it was largely unavailable at the time. Even today, only a few rare Internet users can be expected to have smart card terminals connected to their PCs.

This brings us to another example of a smart card application that experienced massive startup difficulties due to an inadequate infrastructure: electronic purses. Electronic purse cards were issued to nearly the entire population of Germany within a short length of time, and similar actions took place in other countries. Unfortunately, the accompanying infrastructure for making payments and loading the cards lagged behind, with the result that millions of purse cards have remained more or less unused for years. Another unfortunate fact was that the promoters of purse card systems failed to make them usable with public telephones – at that time unquestionably one of the most common applications for chip cards in the form of memory cards.

Another factor with electronic purse systems is that the perceived benefits for users are not especially large. Users must regularly load new electronic money into their purse cards, which is only possible using certain terminals, and they cannot determine how much money is actually stored in their electronic purses without using special technical aids. At the same time, many other cashless payment systems are already in operation, and users have become accustomed to them over the years. Given this background, it is hardly surprising that most electronic purse systems have not achieved significant commercial success.

Systems in which the operator holds some sort of monopoly position – such as companies and government bodies – form an exception to this rule. These operators can simply ignore the wishes of their users when deciding to introduce a smart card system, and they do not have to be afraid that their 'customers' will switch to a competitor. This is shown by quite a few contemporary company card systems and prepaid telephone cards, which were introduced by telephone companies (when they were still state-owned monopolies in Europe) for use with public telephones as a means of payment. Customers had no other choice than to use this means of payment exclusively, even though the cards could not be used anywhere else, as otherwise they could no longer place calls from public phones. However, customers subjected to this form of imposed choice can quickly migrate elsewhere when a competitor arises in such a situation.

In this regard, the SIM cards used in the GSM mobile telecommunication system have proven to be a resounding success. Manufacturers of mobile telephones can produce standard products in large volumes because network operators store their subscriber information exclusively in the SIM cards. Users can also transfer a SIM from one mobile telephone to another one and use the new telephone right away, and the telephone numbers and short messages stored in the smart card will be immediately available in the new telephone. In this way, the three principal participants realize distinct benefits from using smart cards.

The term *acceptance* always implies voluntary use without any form of compulsion. Every operator of a smart card system should bear this in mind from the very start, rather than perhaps thinking that the necessary number of users can be attained with sufficiently persuasive arguments and suitable pressure. The success factors for adequate user acceptance are summarized in Table 8.1. Only monopolists are in a position to be able to ignore these factors, but they should bear in mind that their special status can change in the course of time.

Table 8.1 Principal parameters affecting acceptance of smart card systems as seen from the user perspective, listed in decreasing order of importance. Failure to comply with any one of these parameters can cause a loss of user acceptance

User Parameter	Explanation
Perceived benefit	Users perceive a direct benefit from the smart cards, such as time savings, reduced costs or simplification of previous processes or transactions.
Convenience	The application is convenient for the users, which means their previous customs are taken into account, only minimal actions are necessary on the part of the users, and the entire application is designed to be user-friendly.
Full-coverage infrastructure	A full-coverage infrastructure for using the smart cards is in place before the system is put into service.
Transparency	Users receive complete information about the smart card application and can trust the system. This means that the mechanisms and processes used in the system are understandable to the users, no hidden functions are present, and the usual aspects of security and data protection are respected.
Voluntary nature	Participation in the smart card system is voluntary and other previously used systems remain available at the same time. The users see that their interests are dealt with fairly.
Image of the application	The smart card application does not have a negative image among the general population or in certain groups.

8.2   Tell-tale Signs of Difficult Smart Card Systems

Building a smart card system – no matter how large it is – is a typical project management task, and currently established project development methods can be used to handle it. The types of problems that typically occur in projects are also seen in smart card development, such as insufficient agreement among the involved parties, requirements that are unclear, ambiguous or constantly changing, poor project management, excessively high levels of technical or organizational complexity, political games, and schedules that can ultimately be met only by the combination of a medium-sized miracle and the heroic efforts of a large team.

There are many books about project management that describe how to navigate successfully through these obstacles, which form part of every normal project. However, smart card projects have several specific features that are described in some detail in the following paragraphs. Several tell-tale signs of these problems are also described. If these signs are seen in practice, the project manager should be aware that he must devote special attention to them in project planning and execution.

These signs can also be regarded as critical features that markedly increase the project cost and effort and significantly increase the risk of failure. If at all possible, given the constraints imposed by the requirements of the smart card system, the features described below should be avoided entirely or used only with considerable caution. Experience accumulated over many years shows that over and over again, they have been responsible for massive schedule slips, dramatic cost overruns, and above all quality problems. However, here we would like to emphasize once again that the most important success factor for a project is complete, realistic planning right from the start.

8.2.1   Inappropriate use of smart cards

Considerable difficulties will be encountered as early as the specification phase of a smart card system if the smart cards are intended to be used in a manner that is inappropriate to their principal characteristics. In terms of their characteristics, smart cards are primarily containers for secure data storage and secure program execution. Although they can theoretically be used for other purposes as well, any attempt to do so quickly runs into the limits on their appropriate use.

This can be illustrated by a small example. It involves a manufacturer of switch cabinets, which had the idea of storing all the information about the cabling and electrical operation of each model of a new line of switch cabinets in a smart card located inside the cabinet. The advanced security mechanisms of smart cards would not have been necessary at any point in this project.

The data volume was already several megabytes in the conceptual phase of the project, which meant that it was too large for existing smart cards. However, development of the modular switch cabinet system was expected to take significantly longer than 12 months, so it was assumed that when the time came to deliver the first cabinets, the storage capacity of smart cards available at that time (two generations later) would be sufficiently large. Although the trend in the storage capacity of smart cards conforms roughly to Moore's law, the planners fully overlooked the fact that smart cards are not primarily intended to be used as data storage media, so their development trend does not track the increase in storage capacity of PC memories. Fortunately, this was noticed relatively early in the project and the developers decided to use CD-ROMs instead of smart cards. With a storage capacity of more than 600 MB, CDs provide an adequate margin for future requirements. They are also significantly easier to use in practice and less expensive than smart cards, which are not commonly used in this sector.

8.2.2 Unclear specifications

The right way to formulate specifications for smart cards is described quite extensively in Section 4.5.1. The first visible indication of difficulties with unclear or ambiguous specifications comes when the development team starts asking questions during the implementation phase and they have to be answered in real time. This in turn leads to the real problem, which is that some members of the team may have already finished their work packages without realizing that the specifications were unclear. If critical issues are discovered in the specifications at this point, it can be assumed that some of the components that have already been finished will have to be revised or even redeveloped. This in turn increases costs and leads to schedule delays in the project. Consequently, considerable care should be taken in generating the specifications, and they should be reviewed thoroughly by the involved parties before they are released.

A representative example of an unclear specification is the *Open Platform Specification: Card Specification*[1] published by the Global Platform organization. This document, which fills 200 pages, is a primarily textual description of complex mechanisms for managing applications in smart cards and downloading applications to smart cards. Intensive analysis is necessary to understand this document in detail, and even then it is difficult to be sure that you have understood all aspects of the specification. This document is freely available on the Internet, so you can easily examine its quality for yourself if you are interested. Unfortunately, this specification is the undisputed international standard for managing applications, which makes its structure and form of presentation even more astonishing.

8.2.3 Abundant options

Good technical solutions are characterized by simple, direct implementation of the assigned task. This naturally applies to smart card systems as well. However, uncertainty with regard to the specification or intended use leads to the relatively common practice, especially in the case of applications implemented in smart cards, of providing a large number of options, many of which are kept at a quite general level. This increases development time and complexity and thus increases the likelihood of errors.

For this reason, every option should be examined very carefully before being incorporated in the design in order to ensure that it is actually necessary and useful and not just present because the author of the specification was not sure of what he was doing. In any case, it is always advisable to invest additional time in a thorough analysis instead of introducing new options without careful consideration. In reality, most options are not used later on

---

[1] See Global Platform (2003)

in practice, since nobody is prepared to take the risk of making changes to a working system.

A shining example of a plethora of technically unnecessary options is provided by the ISO/IEC 7816 family of standards. Although these standards are not specification documents, some of their sections address numerous implementation details that certainly have the nature of specifications. For instance, according to ISO/IEC 7816-3, an XOR computation or a CRC computation can be used to generate an error detection code for the T=1 transmission protocol. The latter option has never been used anywhere as yet, and it does not provide any significant advantage in this situation. It only makes the document more complicated. Another example is the nearly inexhaustible range of options for secure data transmission mentioned in ISO/IEC 7816-4. Although they include all conceivable variants of Secure Messaging for data transmission to and from smart cards, deriving useful combinations from the manifold set of options is a nearly impossible undertaking for every user of this standard. Another consequence is that no terminal in the world provides inherent support for this wide range of options without software modifications.

8.2.4   Piggyback applications

The cost of issuing smart cards is unquestionably significant, and there are many applications that could be implemented using smart cards already in the field without any modification to the cards. Probably the most widespread example of this idea is the concept of using German health insurance cards to record flexitime hours in business enterprises. The unique insurance subscriber number stored in each smart card could be put to practical use as a key for administering the actual working hours of employees. This is quite easy from a technical perspective, just like the idea of selling blank smart cards in supermarkets and subsequently loading all the necessary applications into them, which has been floating around for more than 10 years now.

Unfortunately, this idea is impractical because of a large number of contractual and legal provisions. The idea of a creating a secondary use for health insurance cards is simply prohibited by considerations related to data protection legislation,[1] since the data is tied to a particular purpose and cannot be used for a different purpose. Other ideas for piggyback applications regularly fail because of the necessary contractual conditions, which are usually quite tricky.

Some typical issues that must be regulated unambiguously in contractual form are:

- What happens to the data of the piggyback application when the smart card is deactivated or withdrawn for reasons related to the primary application?
- Who is liable for faults in one of the applications in the smart card that are caused by one of the other applications in the card?
- Who pays for the loss of reputation suffered by the original application if the piggyback application exhibits faults that are clearly visible to the user?
- How can the quality and security of the piggyback application be assured?

---

[1] See Section 4.1

All these questions must be addressed, and they result in a rather complex contract and create a considerable risk for the operator of the piggyback application. As smart cards are a relatively inexpensive medium, this normally leads to the decision to issue separate cards for the application concerned instead of using cards already in the field.

8.2.5 Economizing on testing

As is well known, it is not possible to comprehensively test a smart card over the full range of input variables or for all possible combinations of input variables. For this reason, testing is always a form of sample testing, and the adequacy of the tests must be assessed on the basis of their results. However, this also provides a popular incentive for economizing on testing. The risk here is that the operator ends up with software that has not been adequately tested, with the result that faults can occur in the field under the right conditions.

Proof of acceptable test depth can be provided using the metrics of statement coverage and branch coverage. You should attempt to achieve a statement coverage of at least 95% and a branch coverage of at least 80%. In the case of a program with 100 lines, this means that 95 of the statement lines are traversed by the test. Five lines are thus never reached by the test. A branch coverage of 80% means that both decision paths are executed for 80 out of every 100 branches. These metrics are relatively easy to acquire and evaluate, but they do not provide any guarantee that the software is free of defects. Studies have shown that only around 18% of all defects present in software are found with 100% statement coverage, and only around 34% of all defects are found with 100% branch coverage.[1] The ideal situation would be 100% path coverage, but this is not feasible due the amount of time that would be required for testing. It would also not provide a guarantee that the software is free of defects, since some execution paths may be overlooked in the test.

An even handier rule of thumb says that the amount of time spent on generating tests should be approximately the same as the time spent on developing the actual software. Of course, this presumes that the members of the development and test teams have comparable levels of expertise and similar levels of tool support.

There are more than enough examples of scanty testing, and particularly in the PC world it often appears that testing is a totally unknown activity. A good illustration of the effect of scanty testing during software development for smart card microcontrollers is provided by the following example. A small development firm was entrusted with the job of producing an extension to a smart card operating system written in the C programming language. To prepare the developers for their work, several training courses were given before the start of the project. However, smart card operating systems are not designed or constructed to facilitate fast, easy modification by outsiders. This was presumably the main reason why the development fell behind schedule. Unfortunately, the completion date was immovable, so an attempt was made to recover the time lost during development by reducing the amount of time spent on testing at the end.

The delivery date was met in the nip of time and all the involved parties were relieved, at least initially. However, customer acceptance testing revealed a whole series of faults that

---

[1] See Liggesmeyer (2002)

were unacceptable within the planned environment. Subsequent analysis of the statement coverage achieved by the tests showed that it amounted to only 75%. As the software had a scope of around 3000 lines, this meant that 750 lines were not checked at all during testing, even without taking branch coverage or path coverage into consideration. Given this degree of deficiency in test coverage and the presumably hectic development activities due to the schedule delays, it was easy to understand why the quality of the end product was inadequate. Unfortunately, the parties concerned did not realize this until after the fact, when the damage had already been done.

If test coverage had been measured in this project during development, it would have been clear before the acceptance test that there was good reason to assume that rework would be necessary. Given this knowledge, it might even have been possible to shift the allegedly immovable deadline to meet the demands of the situation. In that case, the system could have been put into service faster than by taking the route of first issuing faulty smart cards and then scheduling in a correction loop.

8.2.6   Downloading applications

Marketing types often cherish the dream of freely swapping or modifying application programs in smart cards or activating applications only after the smart cards have been issued. They see this as a means to quickly adapt applications in smart cards that are already in the field to new or altered marketing strategies.

From a technical perspective, modern smart card operating systems and the associated system infrastructure make it easy to download applications, and there are certainly system operators that make extensive use of this technique. A typical example is updating the byte code of microbrowsers in SIM cards already in the field to suit new circumstances. However, large-scale downloading of Java applets, which are packages of program code that is interpreted directly by the smart card operating system, has been used only rarely up to now. One reason for this is the amount of infrastructure bandwidth it requires, and another is the risk that a new or modified application could lead to errors in the smart cards concerned.

Activating functions after the cards have been issued can be rather tricky in some cases. This is illustrated by the following example. Some time ago there was a quite popular code-based smart card application for SIMs that could be downloaded entirely via the air interface. After a major modification to the program flow of the Java applet, a new version of the application was downloaded to the smart cards in the field. One of the consequences of the modification was that a certain part of the application took slightly longer to execute, although this did not appear to have any significant impact. However, the modified timing caused an error with a particular combination of one of the smart card operating systems used in the system and a particular model of mobile telephone, which indirectly resulted in abrupt termination of communication. The mobile telephone only reported a 'SIM Error' message, and the affected users were left wondering whey they were no longer able to place calls. Fortunately, normal communication was restored when the telephone was switched on again, and it remained unimpaired as long as the application in question was not used.

This example clearly shows that even apparently innocent minor changes to applications can lead to considerable damage given an unfavourable combination of circumstances.

For this reason, we strongly recommend that downloaded applications and applications activated after the cards are issued should be subjected to extremely careful testing using a large number of combinations.

8.2.7   Offline systems

Thanks to their characteristics, smart cards are ideally suited for use in systems in which the terminals are not constantly linked to a background system or operate entirely independent of the background system. The principal advantage is that considerable cost savings can be achieved because it is not necessary for all the terminals to have online links. With a suitable system design, such systems can operate flawlessly over long periods of time in a wide variety of use cases.

However, with this sort of arrangement you should always be aware that direct access to the terminals or smart cards from the background system is no longer guaranteed. On top of this, there is a time delay in transferring information to and from the smart cards, with the extent of this delay depending on the system architecture. It can easily amount to several days, or even a few weeks in extreme cases. The consequence is that the transaction problems do not become visible in the background system until after this latency period.

The following example is intended to illustrate the problems that can arise in systems that operate either partially or fully offline. In a relatively large smart card system, changes to the legal framework made it necessary to modify a few bytes in a certain file in all the smart cards in the field. However, this file could only be accessed for writing by the background system after successful authentication using a secret key. The technical implement of this was easy for the terminals that worked online. As soon as a smart card with a file that had not yet been modified was recognized, the terminal established a link to the background system, which then used suitable smart card commands to update the file in a process that was transparent to the intervening network. However, this was not possible with the offline terminals.

It might have been assumed that, sooner or later, all the smart cards would be presented to a terminal capable of operating online. Unfortunately, this was not necessarily true with this particular application. As a result, many of the offline terminals had to be upgraded with modified software by local service personnel. This approach was possible because each of the offline terminals was fitted with a security module that could be used for secure storage of the secret authentication key for data access. Otherwise, the only possible solution would have been to block the use of unmodified cards with the offline terminals, which would have certainly not been a welcome solution for the affected users.

8.2.8   Intolerant smart cards and terminals

Good specifications define all aspects of the interactions between the terminals and the smart cards in detail and in a manner that is not open to interpretation. This is also important in order to avoid interface problems that can otherwise arise quite quickly. However, implementation of the specification must never lead to a situation in which one of the two components actively checks for compliance with the specification and breaks off the interaction if deviations are discovered. What matters in the interaction between smart cards and terminals is that both parties work well together even in borderline cases, rather than discovering any deviations from the documentation. The latter aspect is solely the responsibility of testing. It is an ironclad principle that all components must be as tolerant as possible in their interactions with regard to deviations from specifications.

This can be illustrated by two examples. Some time ago, there was a particular model of terminal used in a smart card system that transmitted the second stop bit of the T=1 transmission protocol outside the tolerance range specified by ISO/IEC 7816-3. The duration of the stop bit was too long. The cause was presumably some sort of imprecision in the implementation of the transmit routine. However, the time intervals between the individual transmitted bits were fully adequate, so error-free data transmission was not a problem. Unfortunately, the receive routine of certain smart cards checked for exact compliance with the specified stop bit duration. As a result, the terminal and the smart cards could not establish a functional communication link. Although the terminal was clearly responsible for this unfortunate situation, this did not help solve the problem, particularly as the terminals were already in operational use and the smart cards had not yet been deployed. The ultimate result was that the tolerance range of the receive routine of the smart card operating system had to be enlarged at considerable expense.

A similarly unpleasant scenario occurred at the application level with a certain type of smart card. It had a file with several unused bits, and the relevant specification stated that all unused bits were to be set to zero. Some time later on, some of these bits were used for an extension to the application. Unfortunately, there was one model of terminal that displayed the message 'Memory error in smart card' in response to this change because it explicitly checked all unused data elements to see whether they were set to zero. In this case, it was possible to load a new version of the software into the affected terminals via remote maintenance, at the cost of a certain amount of time and effort.

8.2.9 Strict compatibility requirements

In relatively small smart card systems, the smart card and terminal components usually originate from a single manufacturer and are designed to work together from a single manufacturer. However, when the system has been in operation for a while and new cards or terminals must be purchased, the question quickly arises as to whether the new components are compatible with the ones already in use. A similar situation exists in larger systems, in which components from several manufacturers or different models of certain components are used right from the start. The GSM mobile telecommunication system provides a good example of an extreme case, in which thousands of different models of mobile telephones with just as many different smart cards must work together flawlessly.

This is only possible with mature specifications and very thorough testing. Nevertheless, problems in the interaction between the components will always occur in certain cases. This includes problems at the electrical level (such as a terminal that does not provide enough current for the smart cards), at the transmission protocol level (such as failure to comply with the timing specifications for data transmission), and typical informatics problems (such as the fact that the integer data type is not supported by some Java cards).

Although compatibility problems can be reduced, they can never be fully eliminated, not even with the best possible specifications and most comprehensive testing.

For this reason, it is advisable to keep the number of different types of components and software versions as small as possible. In the first place, this reduces the number of possible combinations, but it also reduces the likelihood of using a component that can cause problems. This is an example of the axiom that the fewer components a system has, the more reliable it is in operation.

There is also another aspect that must be taken into account. Modern smart cards provide an enormous number of functions and options that can sometimes make certain applications much easier to implement. However, it is not advisable to utilize overly specialized mechanisms of smart card operating systems, since there is always the risk that they will not be supported by a new version of the operating system or a different smart card operating system from another producer. Depending on the situation, this can lead to additional cost and effort to adapt the other components of the system. In principle, this consideration naturally applies not only to the smart cards, but also to all components of the system.

In a North American country, there was a network operator who had a considerable number of Java-programmable SIMs from several manufacturers in the field. For financial reasons, one day the operator decided to take on an additional supplier, which could also provide Java SIMs conforming to the relevant standard. When the first samples of the new SIM were received, the network operator discovered to his surprise that some of the applets, which had already been used for more than two years, generated an error message and thus did not work properly. This naturally led to hurried analysis efforts by the card manufacturer concerned, since it was suspected that there was an error or incompatibility in the operating system. Somewhat to their surprise, after two weeks of research they concluded that the new smart cards were free from errors and all the cards already in the field included an extension to the Java Card API that was not compliant with the specification. The network operator took note of the results of the analysis but was naturally in no position to implement any changes to the smart cards already in the field. The end result of taking on a new smart card manufacturer as a supplier was that the manufacturer made suitable modifications to its operating system to make it compatible with the de facto standard of the existing smart cards.

8.2.10 Excessively stringent security requirements

The most important reason for using smart cards in an informatics system is their ability to protect data and programs against attacks. However, it is quite easy to formulate the security requirements of a smart card operating system or smart card application so stringently that implementation of the requirements is unreasonably difficult or even impossible. Just as with nearly every aspect of a system, you should start by assessing the requirements with an adequate understanding of the situation and issues instead of right away specifying the most extreme solution.

This can be illustrated quite readily by the following example, which involves a company that wished to add a signature function to a company ID card system based on smart cards. The preferred cryptographic algorithm was RSA, and the developers based their selection of the key length on the list of criteria published by the German *Bundesnetz-agentur* (Federal Networks Agency). There they found the recommendation that a key length of 2048 bits is adequate for long-term use, so they incorporated this value directly into the requirements specification. When the time came to select a suitable smart card operating system, they were surprised to discover that although support for this key length was planned for future versions, it was not provided by any currently available system. The reason was that none of the currently available smart card microcontrollers had cryptographic hardware capable of handling a RSA key of this length. This situation nearly led to cancellation of the entire project, as the company began to fear that smart cards were not sufficiently secure.

From a rational viewpoint, this was a rather rash conclusion, as significantly shorter key lengths are still considered to be secure. In addition, such severe security requirements are usually absurd for company ID cards because the likelihood of truly hard attacks is very small with smart card applications of this sort.

8.2.11 Exaggerated future-proofing

Smart cards represent a considerable financial investment when procured in large quantities. Consequently, system operators usually want to see that their investment is adequately secured. This can easily lead them to specify a requirement that the system must be designed to accommodate as many future developments as possible. In many cases, this goes as far as stating that standards still in the draft stage are to be incorporated in the specification. Although such requests are undoubtedly legitimate from the perspective of the system operator, the persons who make these decisions generally do not fully realize the consequences of their requests.

There are two problems that typically arise quite quickly if standards or specifications still in the draft stage must be taken into account in a smart card development project. The first problem is that draft documents are subject to change in various places until they are finalized, which can have considerable impact on the software in the smart cards. The best example of this is a change in a single letter ('a') in the original USIM specification. Instead of specifying the use of an asymmetric encryption algorithm for authentication as originally foreseen, at a relatively late stage it was decided to specify use of a symmetric algorithm. As a consequence, all operating system producers that had generated products in anticipation of the final version of the specification were forced to make extensive changes to their smart card operating systems.

The second problem with referencing draft documents is that they generally have raw edges and are ambiguous or unclear in many places, which naturally leads to the well-known phenomenon of incorrect interpretation during implementation. Incidentally, exactly the same considerations apply to developing smart card software with reference to components (such as terminals) that are still in the development stage.

In the early 1990s, when the essential standard for smart card operating systems (ISO/IEC 7816-4) was still in the draft stage, many card manufacturers attempted to maintain their operating systems as close as possible to the current draft version of the standard. This was based on the rationale that the ability to provide a standard-compliant operating system was a strong sales argument with some customers. However, the experts in the working group responsible for generating the standard (WG4) were not exactly lazy, and they produced new preliminary versions of the standard as often as four times a year. The operating systems based on the standard thus had to be revised each time a new draft version of the standard was released. This sort of after-the-fact development extending over a relatively long time made it nearly impossible to perform effective project planning or effort planning for the operating system development activities. At that time, all smart card microcontrollers had their operating system stored in ROM, so new chips with new ROM masks had to be produced each time major changes were made. Stable versions of the operating systems could only be generated after the final version of the standard was published.

It is certainly possible to incorporate unstable documents or components into a system, and in certain cases it is unquestionably worthwhile, but when making such a decision you should be aware that this can lead to unpredictable and in some cases significant costs for making modifications. You can also calm your fears by remembering that in many cases the issued smart cards will anyhow only have a service life of three or four years, after which they will be replaced by a new generation.

In addition, you should bear in mind that you can never manage to fully predict or anticipate all future aspects, which means that sooner or later compromises will anyhow be necessary. The consequence of all this is that you should focus your strategy on achieving a stable foundation instead of rushing headlong into a risky project by trying to accommodate nebulous future developments.

8.3 Prerequisites for Easy Smart Card Systems

The key factor of any development project that hopes to produce a smart card system that can ultimately be put into service is a clear objective. Experience shows that in relatively large projects, everything that can possibly go wrong usually does go wrong. The art of professional project planning and management is to control the scope of the risks such that a functional smart card system emerges at the end of the project. However, it is certainly possible to create some additional conditions that can make your job significantly easier from the initial phase onwards. You should always take advantage of this possibility. The following subsections describe several important ways to achieve this objective.

8.3.1 Expert advice

People repeatedly generate smart card applications without giving proper consideration to the principal characteristics of smart cards. The results are thus often unnecessarily complicated and make little, if any, use of established mechanisms, and in the worst case they also have security gaps. If at all possible, you should have your application reviewed by experts while it is still in the conceptual design phase. The necessary effort is usually only a few working days, and in return you receive the security of knowing that you are not taking a totally unusual route to your objective.

Alternatively, you can take the approach that has become established over the years with open-source software. This amounts to publishing a complete description of the application, waiting for comments from the professional world, and then incorporating them into the revised application design. After this, you can finalize your design and implement it in the smart cards. Of course, the amount of time required for this process should not be underestimated, and you may not receive any feedback at all for relatively minor applications. In effect, this approach is thus only suitable for relatively large smart card projects.

8.3.2   Foresighted design

Some attributes of smart card applications must be taken into account starting with the conceptual design stage, since they can only be incorporated later on at considerable effort and expense. Probably the best known of these attributes is quality, which must be incorporated in application development as an inherent aspect instead of being planned in as a separate step in the course of the development. The same consideration applies to the attributes of security, data protection, robustness and speed. Attention must be given to these attributes in all steps of the process from the very start, in the sense of performing foresighted application design. If this is overlooked, you can still try to incorporate these attributes in the application later on, but you will always have to accept major compromises if you take this approach.

8.3.3   Prototyping

In many engineering disciplines, it has been common practice for many years to start by building one or more models or prototypes and then base subsequent product de-velopment on experience gained from the prototypes. This sort of approach is rarely used in software development, and when it is used the prototypes are sometimes rashly remodelled into finished products. However, if you take a more professional approach to developing prototypes and incorporate the insights obtained from the prototypes into the development of the final software, this is a powerful tool for purposefully arriving at a mature product.

There is an astonishing variety of advanced software tools available for generating smart card applications. They can be used to generate file-based and code-based applications and load them into sample cards. These cards can then be used to test all typical use cases interactively using a terminal simulator. If it becomes apparent that the software requires some sort of correction, it can be made immediately. After this, you can generate a new sample card and make another pass through the development loop.

Of course, producing a high-quality product involves more than just playing around with a powerful tool long enough to arrive at a working solution. Nevertheless, you can use this method to quickly and thoroughly test the utility of a theoretical design against the most important practical use cases. If this method is used thoroughly and systematically, it is a good way to identify and correct design errors at an early stage. It is particularly convenient because it is by no means time-intensive, thanks to good tool support, and the typical use cases must anyhow be defined in order to ensure adequate test coverage.

As an example, consider a simple smart card application with a few hundred smart carts that was developed for a fitness studio. One of the essential functions of this system was verifying the authenticity of the individual smart cards. The data stored in each smart card consisted of a serial number, a secret authentication key, the name of the customer, and a photo of the customer. Except for the key, this data only had to be read by the terminal, so read access to the corresponding files (with READ BINARY) was allowed and write access (with UPDATE BINARY) was prohibited.

A prototype of the application was loaded into a sample card, and the typical use cases were then tested systematically. During this process, it quickly became evident that an essential use case had been overlooked: the smart card could not be personalized. As only read access to the files was allowed, there was no way to write the personal data of the card user to the card. The design was corrected, and all the use cases were then tested once again. After they could be executed without any problems, an order was placed with a card manufacturer to produce the smart cards. If this mistake had not been found using a sample card, it would not have been possible to use the finished smart cards because there would have been no way to personalize them with individual customer data.

8.3.4 Single-application smart cards

Scenarios in which several smart card applications are bundled in a single multiapplication card are repeatedly proposed in publications. Although realization of this vision has been technically possible for some time now, it has rarely been implemented in practice. This is primarily due to the difficulties associated with the organizational and contractual relationships between the members of the resulting triad: the card owner with the primary application, the application operator with the secondary application, and the card user.[1]

In any case, before deciding for or against a multiapplication card it is advisable to recall that smart cards are actually relatively inexpensive components of the overall system. As arriving at an agreement between the card owner and the application operator is usually a difficult and time-consuming process, it is generally advisable to issue a separate smart card for each application.

8.3.5 Simple structures

Applications in smart cards must have clear, simple structures. This does not mean that they have to be primitive, but instead that they must be easy to understand so that their working mechanisms and security can be assessed by third parties.

Unfortunately, there is still a widely held but mistaken belief that complicated, opaque applications provide protection against attacks. In fact, this is usually the exact reason why attacks are successful, since even insiders cannot understand all the mechanisms of such muddled applications. Ideally, the data and program code of an application should describe it fully and completely, including its static and dynamic aspects. UML has become established as the best way to depict this.

8.3.6 Robust design

Robustness is defined as meaningful behaviour by an entity under conditions that are not addressed (or not fully addressed) by its specification. Robust behaviour of the smart

---

[1] See Section 8.2

cards (and the terminals, of course) can spare the system operator many a situation that would otherwise require modifications to the terminal software or even replacement of the cards.

It is rather difficult to design robustness into an application, since this requires a wealth of experience and a fine sense of judgement. What is important in this regard is that the command sequences between the terminal and the smart card must yield as few unexpected results as possible, as otherwise one of the two parties will terminate the sequence with an error message. As fallback scenarios for responding to unexpected results are normally not envisaged by designers, the user will suddenly be confronted with a 'Card error' message and will be unable to perform the intended transaction. The general advice for achieving robustness is that the smart card or terminal must persist with the transaction as long as possible, but naturally without revealing any secrets in the process.

This can be illustrated by two examples. As a first example, some smart card operating systems allow PIN verification to be disabled by a command (DISABLE VERIFICA-TION REQUIREMENT as specified by ISO/IEC 7816-4 or DISABLE CHV as specified by TS 51.011). If PIN verification has been disabled and a terminal nevertheless initi-ates PIN verification by issuing a VERIFY command, it would be perfectly legitimate for the smart card to respond with a 'Checking error – no specific diagnosis' ('6F00') return code. This would normally cause the terminal to cancel all subsequent steps of the transaction, since the smart card has evidently found an error. The user will be left standing in front of the terminal in puzzlement and will have no choice but to leave without completing the intended transaction. In this situation, a robust smart card appli-cation would respond with a 'Normal processing – Processing performed successfully' return code ('9000'). In this way, the terminal will receive the expected response and can continue with the next command. This does not amount to a security gap, since the user has intentionally disabled PIN verification.

Things are a bit more difficult in the next example, which clearly goes a step further into the grey zone of the specifications. A commonly used method to pass several data items to a smart card with a single command is to structure the data using nested TLV coding. For this purpose, each data item is prefixed by a unique tag and a length parameter. This set of data elements is in turn embedded in a higher-level TLV structure, which is then transferred by the command.

It is well within the realm of possibility for some part of the higher-order TLV structure to be incorrect, such as the length parameter. However, in principle this would not make it impossible to recognize the embedded data and extract it from the TLV structure. The only thing that might cause difficulties would be detecting the end of the overall structure.

A precisely programmed smart card would first check the consistency of all the received length data and immediately perform a hard abort if discrepancies were detected. By contrast, the software of a robust smart card would try to extract all the data elements needed for further processing from the higher-level TLV structure and continue with the rest of the process. As long as the smart card did not reveal any secrets, this would be an entirely acceptable procedure that would help compensate for minor errors in the terminal software. The only essential consideration here is that it must never be possible to use this sort of robust behaviour as a basis for an attack intended to spy out secrets stored in the smart card or discredit the system.

8.3.7 Centralized systems

Although the characteristics of smart cards make them well suited for use in decentralized systems and systems that operate offline, practical experience shows that centralized online systems are distinctly easier to manage. The reason is simply that in such systems the system operator can directly monitor all system components involved in smart card transactions. In addition, changes have to be made in only one location: the background system. Thanks to the steadily increasing networking of our society, it is becoming increasingly economical to operate systems of this sort.

As early as the 1990s, many airline companies introduced centralized systems with contactless or magnetic-stripe customer cards. Around the turn of the millennium, several oil companies and supermarket chains in Germany launched similar systems using magnetic-stripe cards or barcode cards. In such systems, the cards are often used only as identification media and the actual transactions are processed entirely in the background system. Generally speaking, systems with this sort of architecture are only suitable for relatively simple applications such as collecting bonus points, but they offer the advantage of low-risk, flexible operation. As relatively small smart card systems in particular can often be operated as centralized systems without incurring significant limitations, you should keep this approach in mind as an option in the conceptual design stage.

8.3.8 Staged deployment

For security reasons, national payment systems using smart cards are never deployed all at once. Instead, the first stage is a field trial involving at most a few tens of thousands of smart cards and several hundred terminals. The field trial is conducted in a medium-sized city, and it attempts to include most types of businesses in order to cover all common use cases. This approach was used for field trials of the Quick electronic purse system (Austria) conducted in Eisenstadt in 1994, the Mondex electronic purse system (UK) conducted in Swindon in 1995, and the Geldkarte system (Germany) conducted in Ravensburg in 1996.

All transactions are analysed in detail in the background system during the field trial. For this reason, all terminals operate online for the duration of the field trial, even if they are capable of operating offline. Measures for improving the system to correct weaknesses discovered in the course of the field trial are also formulated during the field trial stage. These weaknesses may lie in the timing of the smart cards and terminals, for example, or in the number of transaction data sets arriving at the background system.

During a field trial several years ago, the testers discovered that a large number of cancelled transactions occurred with a particular model of terminal. An attack was suspected, but detailed analysis showed that in each case the user pulled the card out of the terminal during the transaction. The underlying reason was a communication problem between the terminal and the smart cards that drastically prolonged the data transfer time.

The problem was corrected by a quickly implemented modification to the transmission protocol in the terminal. This correction was implemented in a 'sneaker update' action by a team of service technicians of the terminal manufacturer, who travelled from one terminal to the next in the city and loaded new software into the faulty terminals.

The information necessary for initiating the next stage, which is national deployment of the payment system, can be collected in this manner within a few months. However, it is important to allow sufficient time to implement modifications (including major changes) and corrections to all the system components before national deployment. The required interval can easily be as much as one year.

This procedure for introducing a large payment system, which has been used successfully for many years, can serve as a good model for smaller smart card systems. Following completion of the prototype phase, which is restricted to the laboratory environment, the next phase is to issue a relatively small number of smart cards so that they can be used by trial users for a certain length of time in typical use cases. If necessary, these smart cards can be collected at the end of the field trial and new cards can be issued before the start of normal system operation. This would be a reasonable approach with minimal risk. In any case, you should avoid a 'big-bang' system launch, since it entails a considerable risk of being forced to withdraw the system from operation if difficulties arise and then relaunching it after the necessary corrections have been implemented.

8.4   In-field Faults

The biggest problem that can occur with a smart card application is a software or hardware fault in the cards in the field. This is a critical problem because, unlike the situation in the PC world, there is almost no way to update the software in the smart cards after they have been issued. Smart cards are similar to hardware components in this regard, since they also cannot be modified after product delivery. The available options for dealing with a fault in the field are rather limited, as generally speaking even the system operator no longer has direct access the cards that have been issued. As the number of smart cards is also quite large in most cases, the scope of the loss can be correspondingly great.

8.4.1   Fault classification

The causes of faults in smart cards can be traced back to either the hardware of the microcontroller or the software of the smart card application or operating system. Of course, faults arising from the interface between the hardware and the software also occur in practice, but they are usually difficult to classify.

Another attribute of a fault is its mode of occurrence. Some faults are deterministic, which means that they occur consistently under certain precisely defined conditions. By contrast, probabilistic faults either occur only under highly complex conditions or appear to be random, so they are difficult to predict. Such faults are never truly random, since smart cards are strictly deterministic at the processor level and do not have true physical random number generators. An example of a fault with a probabilistic nature would be a smart card operating system crash due to a stack overflow that depends on the sequence of use of the five most recent applications.

Another factor that must be considered is whether the fault is permanent or temporary. A permanent fault persists after the card has been reset by a warm reset (with power still applied) or a cold reset (executed during the smart card power-up sequence). A temporary fault may disappear after a reset or some other specific event, only to reappear later on after the next trigger event.

In the case of software faults in smart cards, a distinction can be made between latent faults and acute faults. Latent faults are known to the card manufacturer or system operator but have not yet occurred in practice. The reason for this may be that the function concerned has not yet been put into service or is used only rarely by the cardholders. By contrast, acute faults are faults that have already occurred in smart cards in the field and have a detrimental effect on system operation.

Table 8.2 summarizes the attributes that can be used to classify smart card faults.

Table 8.2   A useful classification scheme for smart card faults

Cause	Hardware and/or software
Reproducibility	Deterministic or probabilistic
Persistence	Permanent or temporary
Visibility	Latent or acute
Impact	Function and/or security
Probability of occurrence	Percentage figure relative to lifetime or per year

Depending on the application, it is also of considerable importance whether the fault affects system operation (functional fault) or system security (security fault). There are unquestionably applications in which the occurrence of a functional fault represents a small catastrophe for the system operator but a minor security leak is only a cosmetic error. However, in most cases security faults are more critical than functional faults, and they demand a response that is as quick as possible and, depending on the circumstances, suitably discreet.

Faults can also be classified in terms of their probability of occurrence. This is typically expressed relative to the planned service life of the smart cards, so the probability of occurrence of a particular fault might be stated as 30% over a period of three years. If the service life is not defined, the probability of occurrence of a fault can also be expressed relative to an interval of one year.

8.4.2   Fault impact

If a fault occurs in a smart card after it has been issued, the measures taken to deal with it depend primarily on the impact of the fault. The simplest approach is to identify a guilty party and present it with a claim for damages. However, this will not eliminate the actual problem, especially since it is usually only possible to claim compensation for direct losses. Compensation for indirect losses, such as loss of reputation or lost profit, is usually excluded by the provisions of the relevant supply contracts.

For this reason, For this reason, the most intelligent action after you determine the cause of the fault is to determine the impact of the fault. This is because under certain circumstances, it may be possible to take suitable actions to compensate for the fault to such a degree that the card users hardly notice the fault or are totally unaware of it. This is much better than taking the uncertain route of pursuing a claim for damages, at the end of which the actual problem will still be unresolved.

Naturally, faults in smart cards do not occur only in the field, but also during the analysis phase of software development. The incurred cost of faults that occur during the analysis, design, coding and testing phases is typically of the same order of magnitude as the cost of the actual software development.[1] However, the costs of faults that occur in the field are distinctly higher than with other forms of software, since the options for accessing the smart cards issued to the users are naturally very limited. This is illustrated in Table 8.3 with some representative examples for a medium-sized smart card system.

Table 8.3   Estimated costs of typical faults that can occur in a smart card system during operation. The calculated losses are based on an assumed card price of 5 euros

Type of Fault	Action	Cost
Secondary card function unavailable	Retroactive price discount	10 000 smart cards, 0.5 euro retroactive discount per smart card; direct loss: 5000 euros 100 000 smart cards, 0.5 euro retroactive discount per smart card; direct loss: 50 000 euros
Primary card function unavailable	Card replacement initiated by user	10 000 smart cards, replacement cards provided at 5 euros per smart card; direct loss: 50 000 euros 100 000 smart cards, replacement cards provided at 5 euros per smart card; direct loss: 500 000 euros
Total card failure	Card replacement initiated by system operator	10 000 smart cards, replacement cards provided at 10 euros per smart card; direct loss: 100 000 euros 100 000 smart cards, replacement cards provided at 10 euros per smart card; direct loss: 1 000 000 euros

The impact of a fault on the user depends strongly on the actual malfunction. If a secondary function of the smart card no longer works because of a fault, the user can still use the essential functions and the impairment of use of the card is limited. Wholesale card replacement is thus generally not necessary in case of a fault in a secondary function.

The situation is different with faults that affect the primary functions of smart cards. In such cases, the user is strongly impacted and replacement of the smart cards on the initiative of the system operator can be appropriate if the smart card has only a few

---

[1] See Boehm (1981)

primary functions or just one. In the worst-case situation where the fault renders the card entirely useless, card replacement is the only possible solution.

This can be further illustrated using SIM cards as an example. A possible secondary function with SIM cards would be an information service implemented by an application in the smart card that regularly provides the user with current information, such as the latest football scores, by means of text messages. Although failure of this function might be annoying for some people, it would be an acceptable impairment for most users. People who were seriously upset by incorrect operation of this information service could also personally take the initiative to have the card replaced at a service centre of the network operator. By contrast, if one of the primary functions (such as placing telephone calls) were to fail, this impairment would unquestionably be unacceptable to the users and the network operator would take the initiative to launch a recall campaign for the affected cards. The same response must be taken if the smart card fails entirely due to a fault. Table 8.4 summarizes the actions to be taken according to the probability of occurrence of a fault.

Table 8.4 Fault impacts and corresponding typical actions by system operators

Fault Impact	Probability of Occurrence	Typical Action
Secondary function unavailable	Low	No action
	High	Card replacement initiated by user
Primary function unavailable	Low	Card replacement initiated by user
	High	Card replacement initiated by system operator
Card failure	Low	Card replacement initiated by user
	High	Card replacement initiated by system operator

If a fault occurs only sporadically, the necessary measures depend strongly on the probability of occurrence. If the probability of occurrence is sufficiently low, the operator may decide not to launch a replacement campaign even if the fault results in total failure of the smart card, since such campaigns always have a negative impact on the reputation of the system or application. The probability can be computed from the number of cards subject to the fault and the time frame over which the fault can occur. Specific probability figures must be calculated individually for each smart card application concerned. After this, an assessment must be made to set the threshold level for intervention by the system operator.

8.4.3 Actions in response to a fault

The basic rule when a fault occurs is 'don't panic'. You should always bear in mind that there is surely enough time to perform a systematic analysis and to initiate and implement the appropriate measures.

If a large number of smart cards in the field are affected, the potential loss can be considerable depending on the circumstances, which usually leads to a high level of interest from management and hectic activities on the part of the investigation team. It is therefore advisable to prepare an action plan for dealing with faults – and to do this before any fault actually occurs. Marking escape routes in buildings can serve as a good example here: you don't wait until a fire breaks out to mark them, but instead mark them in advance. The same consideration applies to preparing an action plan for handling in-field faults in smart card systems, which you naturally hope will never have to be used.

If a fault nevertheless occurs, you can use the action plan to guide your response. An important principle here is that the first priority should be determining the cause of the fault instead of looking for the guilty party. Systematic, purposeful investigation of the cause of the fault, independent of the question of who is responsible for it, is highly important, since otherwise it is easily possible for the underlying cause of the fault to remain undiscovered, with the consequence that the same misfortune can occur again sometime.

When putting together a search team, you should select a few highly experienced employees with a combination of hardware and software competence to look after this task. You should intentionally keep the size of the team small, and you will need a competent team leader to direct the team. You can be guided here by organizational structures commonly used for dealing with crisis situations. The team leader must have an extensive scope of authority, and he or she also acts as the interface to the other parties concerned.

All members of the team must devote their attention exclusively to the fault, and they must have direct access to the infrastructures and employees of the involved companies as necessary for investigating the fault. This is the only way to ensure that the cause of the fault can be found as quickly as possible and perhaps even that a workaround solution can be generated. All outsiders and interested parties (including management) must be consistently excluded so that the team members can do their work quickly and effectively.

If the cause of the fault can finally be found, it must always be confirmed by a second, independent analysis. Incorrect assumptions about faults lead to a loss of confidence in the search team, which must be avoided at all costs. Consequently, the first fault analysis that is communicated must have a high probability of being correct. The 'salami tactic' of releasing results piecewise, no matter how appropriate it may be in other situations, only leads to wild assumptions and rumours in this situation and does not aid the investigation of the fault. Consequently, this tactic should be avoided.

The search usually ends with the generation of a report by the search team. This report contains a description of the fault, an exact depiction of the cause of the fault, proposed measures for correcting the fault, and recommendations for ways to avoid or prevent the fault in the future.

8.4.4   Fault search procedure

The basic fault search procedure consists of formulating a set of hypotheses based on the described symptoms of the fault. As is well known, a hypothesis is a self-consistent but unproven assumption that can be confirmed or rejected by experimental means. After a set of related hypotheses – the hypothesis tree – for the cause of the fault has been generated, the next step is for the search team to formulate the experiments necessary to confirm or reject each of the hypotheses. The following step is to start performing the experiments.

Depending on the results of the individual experiments, it may be necessary to investigate additional details by means of further experiments. Table 8.5 shows part of a typical hypothesis tree, along with the associated experiments and results for an assumed fault. Hypothesis trees of this sort can easily encompass 30 to 60 hypotheses in case of a complex fault, and they must all be tested experimentally.

Table 8.5   A hypothesis tree for fault-finding in a smart card system. The observable fault was sporadic crashing of the smart card operating system during command processing. The cause ultimately proved to be an excessively long wire in the supply voltage circuit of the terminal. It caused a brief sag in the voltage supply during intensive EEPROM write accesses in the smart card, which quite properly triggered the undervoltage detector of the smart card

	Hypothesis		**Experiment for Testing the Hypothesis**	**Result**
1.	Software fault in the smart card operating system	1.1	Test the smart card operating system in a microcontroller simulator under the conditions that cause the fault to occur	Not confirmed
2.	Sporadic RAM fault	2.1	Write a data pattern to RAM, read it immediately, and compare it with the written pattern	Not confirmed
3.	Voltage detectors too sensitive	3.1	Provisionally solder a capacitor between the supply line and ground directly at the module contacts	Confirmed
		3.2	Use a different model of terminal	Not confirmed
4.	Supply voltage dropouts	4.1	Use a multimeter to measure the supply voltage	Not confirmed
		4.2	Use a high-resolution digital storage oscilloscope to measure the supply voltage during EEPROM write accesses	Confirmed
	. . .		. . .	. . .

The key factor in searching for a fault is the ability to reproduce the fault in an isolated environment. For this, you require at least one of the affected smart cards and possibly also a suitable terminal if this is necessary to reproduce the fault. Depending on the situation, you may also need a link to the background system or a simulated background system. In addition, you must determine the version numbers of the smart card operating system, application and personalization. If the smart card is still operational, this information can often be read out using simple structured commands such as GET DATA.

Of course, you also need a detailed description of the fault and the conditions under which it occurs, which preferably should be provided by the person or party that discovered the problem. It is also advisable to take possible side effects of the fault directly into account in your investigation.

Determining the cause of a fault can be rather difficult and time consuming, since smart cards are especially well armoured against all forms of external analysis. Searching for a fault can take as little as a few hours or as much as 10 to 15 working days. The latter is primarily the case with faults located in the interface between the hardware and the software.

The first step in searching for a fault should consist of using a suitable tool to log the communications between the terminal and the smart card. You can also use a digital storage oscilloscope to record the communication, which has the advantage that it captures the signal processes at the electrical level. This is especially useful in case of problems with the smart card hardware. The data obtained in this way can be used as the basis for all further analyses. In some cases, it can also be helpful to measure the current drawn by the smart card microcontroller during operation in order to identify EEPROM write and erase accesses. In combination with the communications and the source code of the software used in the smart card, this provides quite useful information about program execution.

If the fault is located in the software of the smart card, you can also employ commonly available development tools for smart card software, such as emulators and simulators. However, this requires intensive support from the producer of the operating system, since access to the entire source code is essential for this approach. You can attempt to track down pure software faults by using a simulator, although this has the disadvantage that it does not reproduce real-time behaviour. By contrast, an emulator uses hardware to provide a nearly perfect imitation of a real smart card microcontroller. For fault-finding, this has the enormous advantage that the system exhibits real-time behaviour and you can set breakpoints and read out the contents of all the memories of the (emulated) smart card whenever required. Emulators are the most powerful tools for fault-finding.

If you suspect a fault in a Java Card applet, you can make good use of the simulation tools of the applet development environment. You should also use these tools first to the extent that it makes sense to do so, as using simulators and emulators is significantly more time- and labour-intensive.

Once you have finally identified the fault and verified the assumed cause using at least one different experiment, it is also advisable to fully clarify the fault mechanism at the theoretical level. Only then can you be sure that you have truly found the fault and determined all the causes. This is important for avoiding or preventing the fault in the future. Finally, you have to determine which cards are actually affected by the fault. This can be done using production and delivery data, and it forms the basis for all further actions.

8.4.5 Fault remedies

After the fault mechanism and the number of affected smart cards have been determined, you can start exploring the options for remedying the fault. It helps to involve the members of the search team in this process, since they have the best understanding of the cause and impact of the fault. Depending on the situation, it is also advisable to call on other experts as well in order to arrive at a realistic and effective remedy to the fault. Always bear in mind that all actions for correcting or eliminating faults must be performed as discreetly as possible. Nobody benefits from correcting a fault in a smart card system with lots of fanfare and publicity. In the end, such an approach only damages the reputation of the system.

Various options are available, depending on the specific fault. With some faults, the probability of occurrence can be reduced or even eliminated by changing certain parameters in the smart cards. Depending on the application, it may even be possible to do this without the users being aware of it. However, in some cases it is necessary to disable a particular function of the smart cards. If this does not involve one of the primary functions of the smart cards, it will normally be accepted by the card users without a major protest. Here it helps that many consumers have come to accept software defects as an unavoidable fact of life, despite the fact that this is actually a shameful situation for the entire industry.

On the other hand, if a primary function of the smart cards is impacted by the fault or the smart cards cannot be used at all after the fault occurs, there is no alternative to card replacement. Nevertheless, this must be done with suitable intelligence and sensitivity, rather than with brute force. If only a small number of cards are involved, you can selectively request the affected card users to exchange their cards as necessary. This approach is also suitable if the fault can be remedied by simply downloading a new version of the faulty application from a special terminal. Corrections to parameters or applications can also be made within the context of normal transactions if this is permitted by the system infrastructure. However, this requires upgrading the terminal software to expand its functionality.

In the case of pure online systems, this sort of correction can even be carried out directly from the background system, assuming that it can be upgraded for this purpose at an acceptable cost. However, only certain types of faults – primarily those in the application area – can be corrected in the field using these methods. It is usually not possible to use downloaded software to correct problems in the operating system or the smart card hardware. In such cases, there is no alternative to card replacement in case of a serious fault. A similar situation exists when a large number of cards in the field are affected by a fault. In this case the general tendency is to replace the cards, despite the considerable loss of reputation this entails, instead of attempting to modify the software in the smart cards.

Chapter 9

Illustrative Use Cases

This chapter contains descriptions of several smart card applications of widely varying nature. What they all have in common is that they involve a small to medium number of cards. Such applications are often considerably more interesting than well-known large applications such as mobile telecommunication and payment systems, since large systems are driven primarily by aspects such as secure system operation and compatibility over card generations and usually cannot afford to take innovative approaches.

The examples described here do not always illustrate perfectly executed solutions – in fact, some of them show distinct weaknesses. These examples have been chosen quite intentionally, since it is often possible to learn more from a badly implemented solution than from one that is implemented properly. Of course, you can learn the most by keeping your eyes open and trying to understand every smart card system you encounter as well as possible.

9.1 Monastery Card

On a splendid summer day, I was invited to one of these typical events organized for developers and other creative types so they can discuss technical innovations and new business opportunities. The workshop was held in a former monastery, which occupied almost the entire surface of an island in a small lake. The fortress-like monastery building could only be reached on foot by a narrow bridge.

At the reception desk, the participants were given smart cards for access to their rooms. I found this quite interesting from a technical perspective, since up to then I had only seen normal keys and the usual improvised systems with primitive punched cards or magnetic-stripe cards. After I found my room in the old building with its metre-thick walls and opened the ancient oak door using the smart card, I unpacked my rucksack.

As I still had some time before the start of the workshop and I was curious about the smart card, I tried to learn a bit more about it. I expected it to be a microcontroller smart card that would perform some sort of authentication with the terminal in the

Smart Card Applications: Design Models for using and programming smart cards   W. Rankl
© 2007 John Wiley & Sons, Ltd

door lock. If the authentication was successful, the terminal would grant access to the room. The amount of energy needed for this is so small that a battery could power a lock of this sort for several years. Incidentally, relatively sophisticated systems do not even need a battery, since the necessary energy can be generated by actuating the door handle.

I thus inserted the smart card in the card reader of my notebook computer and initiated a standard activation sequence for microprocessor cards. You can imagine my surprise when I discovered after a few experiments that the card was actually a memory card with an I^2C bus interface. Such a card is nothing more than a portable data storage medium, similar to a magnetic-stripe card, and usually unable to perform any reasonable sort of authentication. With my card reader, I could read the data content of the card using a standard read command (READ BINARY). An analysis of this data showed that the card had 256 bytes of freely readable memory, but only a few bytes were actually used. A portion of the memory content is shown in Table 9.1.

Table 9.1 Part of the contents of a memory card used in an access control system for hotel rooms. The memory content is shown in hexadecimal form in the upper line of each row, with the ASCII equivalent in the lower line as appropriate

	x0	x1	x2	x3	x4	x5	x6	x7	x8	x9
0y	...	...	...	...	...	...	...	...	...	...
1y	'32'	'34'	'37'	'00'	'00'	'00'	'00'	'00'	'00'	'00'
	"2"	"4"	"7"							
2y	'00'	'00'	'00'	'00'	'00'	'00'	'00'	'00'	'00'	'00'
3y	...	...	...	...	...	...	...	...	...	...

As I examined the content of the memory a bit more closely, I thought to myself: 'This is hardly a technically sophisticated solution'. Aside from the obligatory ATR, only 3 bytes at the beginning of the memory were used, with all the rest being set to '00'. I was rather disappointed when I realized that these 3 bytes contained the ASCII equivalent of my room number.

With that discovery, it became obvious that the door terminal simply read part of the memory of the card and compared the data it obtained with the room number stored in the terminal. If the two numbers matched, the door could be opened, while otherwise it remained locked. The security of this system is based on the assumption that the technology used is not generally available, but this has not been the case with smart cards for many years already.

Armed with this knowledge, my next logical step was to try a little experiment, especially since another workshop participant had now checked in to the room next door. The smart card did not have any write protection, so it was easy to change the room number to a different one. All that was necessary was to instruct the card reader to issue an UPDATE BINARY command to replace the room number in my card with the room number of the other guest. Of course, I asked him before doing this, since I did not want to be mistaken for a burglar. After the number was changed, we both tried the modified card in the door of the neighbouring room. It worked perfectly.

The contrast between the room with its ancient, metre-thick walls and robust door and this rather amateurish access system can best be compared to a safe constructed with a back wall made from cardboard instead of heavy steel plate. As long as you do not see the back of the safe, the overall system makes a robust, well-conceived impression, but you will have to revise this impression after you see behind the facade.

Improved solution: cryptographic checksum It would be rather easy to improve the security of this access system somewhat. The simplest and least expensive approach would be to secure the stored room number with a cryptographic checksum (CCS) and use memory cards that require PIN verification before allowing write access to the memory. In this case, the door terminal would only have to re-compute the cryptographic checksum after reading the data from the card and then compare it with the checksum read from the card. Access would be granted if the two checksums were the same, while otherwise the door would remain locked. Write protection using a PIN code would prevent outsiders from modifying the data in the cards used in the system. At minimum, this would force potential attackers to bring along their own cards.

This approach to improving the system would not provide any protection against cloned cards, since data could still be read from such cards, but it would protect against the simplest forms of attack. Extending the functionality of the door terminal to include computing and comparing cryptographic checksums would not present any technical difficulties, since it only requires adding the necessary functions to the software in the microcontroller. The security of an access system modified in this manner would be at least as good as that of a system using conventional mechanical locks.

Improved solution: processor card A much better technical solution would be to use processor cards and slightly more powerful terminals. In this case, the terminals would perform a challenge–response authentication as shown in Sequence Chart 9.1, using a different key for each door. The first step of this authentication consists of the terminal sending a random number to the smart card. The card uses its key to encrypt the random number and then returns the result to the terminal. Next, the door terminal checks whether the result provided by the card matches the result calculate by the door terminal. If it does, the door is unlocked.

Terminal (IFD)		Smart Card (ICC)
INTERNAL AUTHENTICATE (RND)	$\longrightarrow$	$X = E(K_{ICC}, RND)$
	$\longleftarrow$	Response [X]
$X' = E(K_{IFD}, RND)$ IF $(X = X')$ THEN access allowed ELSE access not allowed		

Sequence Chart 9.1 A challenge–response authentication between a door terminal and a smart card, as described in the text as an improved solution for a room access control system

The main advantage of this approach to improving the system is that access permission can be checked using a single command. One difficulty with this approach is that it creates a problem if it is necessary to have a card that can be used to unlock all the doors, similar to a master key in a conventional lock system. However, this requirement could be met by having the door terminal perform additional authentication attempts if the first authentication fails. Alternatively, keys that permit access to more than room could be incorporated in the system.

A variation on the above solution is to use mutual authentication of the door terminal and the card with a shared key. If this authentication is successful, the smart card grants the door terminal read access to a file containing a list of all room numbers for which it has been granted access permission. This makes it possible to produce staff cards that can be used to access multiple rooms or even all rooms.

These improvement options require converting the door terminals to use a processor card protocol (T=0 or T=1) and providing them with sufficient computing power to perform encryption. Processor cards are slightly more expensive than memory cards, but they would not need special programming in this case because it is a purely file-based application. Such a system would be significantly more secure and more flexible than the original system or a conventional lock system.

9.2 Access Card

It is usually quite entertaining to critically examine the high-security access control systems portrayed in films and mentally picture their operating principles. The scenarios in such cases are generally quite similar and are based on the premise that the protagonist wishes to gain access to a high-security area by normal means, which means without direct destruction of the door or lock system. A card is used for this purpose. This is something that warms the heart of anyone with an interest in smart cards, since it can be taken as evidence that general awareness of the advantages of smart cards has reached as far as the realm of the film industry.

Unfortunately, the course of events usually takes the following turn: first, the protagonist grasps the card by its long edge, approaches a card reader with a fiercely determined expression, and pulls the card vigorously through the slot of the reader. A signal then sounds, and the door can be opened. This spoils the effect in a single blow for a smart card expert in the audience, who at best can only manage a condescending smile. In most cases, the card expert then attempts to explain to his neighbours in the audience why the portrayed form of access control is much better suited to protecting empty wooden sheds than high-security areas.

The scene described above clearly indicates that the protagonist is using a magnetic-stripe card. This can be seen from the fact that the card is pulled through a slot. By contrast, a contact smart card would be inserted in a terminal or a contactless smart card would be held next to an antenna surface. The access control system in the film is not secure because the data content of a magnetic-stripe card can be copied quite quickly using simple equipment. The access control terminal cannot tell whether it has read the magnetic stripe of a genuine card or a copied card. The result is an illusion of security instead of effective protection against unauthorized access to an area.

A variety of smart card applications and associated infrastructures that can be used to achieve a relatively high level of security are described below. The next time you have to sit through a film scene like the one described above, you will thus be in a position to casually outline a better solution and thus present yourself to your fellow viewers as someone who knows a thing or two about smart cards. Of course, before you launch into an explanation it is highly advisable to ensure that the moment is suitable and the other viewers are actually interested in such details.

An access control system requires terminals, either located close to the doors or built into the doors, that can release the door opening mechanisms. The terminals can be operated either offline or online with respect to a central background system. The principal advantage of online terminals is the ability to block specific smart cards on short notice. However, online links generate slightly higher operating costs. Nevertheless, widespread availability of wireless technology and networks has reduced these additional costs significantly compared with the situation a few years ago, when separately installed lines were often necessary. Offline terminals can operate independently, which means they do not need a high-availability network infrastructure linked to a background system. However, it is not possible to block smart cards directly in an offline system, so manually updated blacklists must be stored in the terminals of such systems.

Mixed forms are sometimes used in practice, consisting of terminals that are normally used online but can also be used offline. However, in this case care must be taken to ensure that no security gaps arise from the fact that the smart cards used with terminals operating in offline mode cannot be blocked directly. Even with modest equipment, it is not difficult for an attacker to disrupt the operation of a wireless network or cable network to such an extent that the terminals are unable to establish a link to the head-end system.

An access control system based on smart cards has clear advantages relative to a lock system using conventional keys. For instance, the access rules can be formulated quite flexibly and can be modified easily. Administration of the access rules is also easier than in systems in which normal keys are issued to users, and lost-card situations can be handled quickly and at a reasonable cost. It is also easy to configure access privileges for administrative functions, such as cards for building caretakers and janitorial staff. In many cases, the net advantages relative to a conventional lock system justify the additional investment in smart cards, terminals and installation. This sort of solution has been used for many years already in major hotels, such as the system described in Section 9.1.

From an informatics perspective, one of the prerequisites for any access control system is clear identification of every access point. The following examples are based on a scheme widely used on an international scale, in which the floor number and room number form a three-digit number. A minus sign can be placed in front of the floor number to address floors below the ground floor. The ground floor is assigned the number '0', which is common practice in many European countries.

The coding scheme must be designed such that the room numbers can be processed by an automated system. Here it's necessary to make a trade-off between storage requirements and ease of use. In this case a BCD scheme is used, so two digits of the number can be stored in a single byte. The value 'A', which is not used for any digit, is used to represent the minus sign, and a full stop (coded as 'B') is used as a separator between the floor number and the room number. The value 'C' is designated as a placeholder, which can be used in any individual position. The data elements are arranged left-aligned and padded with a trailing 'F' code as necessary to end on a byte boundary.

In this scheme, room 5.13 is located on the fifth floor and has room number 13. The coded value for this is '5B13'. The room with number −1.01 is located on the first floor below the ground floor and has room number 1. It is coded as 'A1B01F'. Although this system has distinct limits for use in large buildings, it is quite suitable for use as an illustrative example. It can also be extended quite easily for use with larger systems.

The infrastructure of this access control system has the following characteristics. First, the terminals do not have keypads, since they would be too expensive and it would take too much time to gain access to the rooms. Ideally, the terminals have sufficient memory to store blacklists of specific cards, which are manually loaded into the terminals. The terminals can be used either online or offline, depending on the specific circumstances, as illustrated conceptually in Figure 9.1.

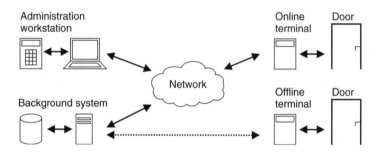

Figure 9.1 Basic architecture of an access control system using smart cards and terminals that can operate either online or offline

The smart cards used in this system are intended to be used for file-based applications,[1] which means they do not necessarily have to support downloadable program code. The system employs a conventional file system and standard commands as specified by ISO/IEC 7816-4. As long as it is not necessary to manage hundreds of access points, smart cards with a few hundred bytes of EEPROM will be adequate for this application. Naturally, this application can also be generated entirely in Java, and it can be used with either contact cards or contactless cards.

The general principle of the access control system is based on conventional lock systems with keys. Such systems have keys for individual locks, keys that can be used to operate all locks on a particular floor and master keys that can be used to operate all the locks in a building.

If this concept is transferred to a smart card system, it means that a unique room number must be stored in each card for each point where access is granted to the card holder. If a card provides access to several rooms, the corresponding room numbers must be stored in the card. This means that the full set of room numbers would have to be stored in a smart card for a building caretaker, which would be somewhat cumbersome. A more

---
[1] See Section 3.3.2

intelligent approach would be to use placeholders for the room numbers so all rooms on a particular floor or all the rooms of a building can be addressed using a single entry.

The access information could be stored in each smart card in an EF file or a data object (DO). In this case, we chose to use data objects for storage. This has the advantage that each access terminal can request the relevant access information directly from the smart card. With a file-based application, the terminal would have to first search for the data in a file, which would take somewhat longer. However, it would certainly be possible to store the data in an EF with a transparent structure instead of using data objects. All data for this application is contained in a directory (DF) with its own AID, so the application can also be used in a multiapplication card.

To enable the system to address each smart card individually, it is a good idea to include a card number or ICCID (integrated chip card identifier), which corresponds to the key number in a conventional lock system. The card numbers also serve as indices for managing all the access privileges, and they can be used to generate the blacklist. This application uses secret keys for authentication of the terminals and smart cards. To make key selection as simple as possible, the keys are selected implicitly by the authentication commands. Although this means that the keys cannot be selected directly by the terminals, it does not impose any restrictions on the functionality of the access control system.

If the data objects could be read freely, it would be quite easy for an attacker to clone the access cards. The attacker could simply read out all the data from a card and then write it to a new card, which of course would have to support the necessary commands and data object management functions. However, this is readily possible with commercially available Java cards.

Consequently, we add an access condition to prevent free readout of the smart card. In this case, the condition is authentication of the terminal by the smart card, so the data in the smart card can only be read after the card has determined that the opposite party is a genuine terminal. As the terminal must also know whether the smart card is genuine, it is convenient to perform both authentications in the same step.

The appropriate command for this purpose is MUTUAL AUTHENTICATION, and it is defined in the ISO/IEC 7816-4 standard. The terminal and smart card need a secret key for mutual authentication. This key is also stored in a data object.

Table 9.2 and Table 9.3 list the data objects required by the access control application. They are stored in two EF files using TLV data structures. Beside the data and keys for operational use, another key is necessary for administrative purposes. It is used to authenticate the terminal in order to attain a state in which all data objects can be written.

Sequence Chart 9.2 shows the usual command sequence for checking access privileges during operational use. The first step of this sequence consists of initiating mutual authentication of the terminal and the smart card. In this process, each of the communication partners checks whether the other one is genuine. If the result is positive, the smart card knows that the terminal is genuine and it grants the terminal access to the stored access data. In the other direction, the terminal also knows that the smart card is genuine and thus that the data received from the smart card in the following step is authentic. The terminal requests this data from the card by passing the correspond tags as parameters. If the associated data is stored in the card, it is returned to the terminal in response;

Table 9.2   Data objects and associated access conditions for an access control application. The data is stored in an EF, which has corresponding access privileges for operational and administrative use

File:	EF$_{\text{AccessData}}$		
FID: SFI:	'EF01' '01'		
Structure: File size:	TLV data structure ...		
Access conditions:			
	READ: UPDATE:	Mutual authentication of the terminal and smart card Unilateral authentication of the terminal by the smart card	
Tag	Length (bytes)	Value	Description
81	4	...	Card number (ICCID)
83	2	'5B13'	Data object for room 5.13
84	3	'A1B01F'	Data object for room −1.01
85	2	'5BCC'	Data object for rooms 5.00 to 5.99
...		...	

Table 9.3   Keys and associated access conditions for an access control application. The keys are stored in an EF, and they can be used directly by the corresponding commands in normal operation. Suitable access privileges are defined for administrative purposes

File:	EF$_{\text{Key}}$		
FID: SFI:	'EF02' '02'		
Structure: File size:	TLV data structure 20 bytes		
Access conditions:			
	READ: UPDATE:	Never Unilateral authentication of the terminal by the smart card	
Tag	Length (bytes)	Value	Description
82	8	'57 6F 6C 66 67 61 6E 67'	8-byte DES key for MUTUAL AUTHENTICATE
83	8	'56 69 6B 74 6F 72 69 61'	8-byte DES key for EXTERNAL AUTHENTICATE

Terminal (IFD)		Smart Card (ICC)
SELECT [AID]	$\longrightarrow$	
	$\longleftarrow$	Response
GET CHALLENGE	$\longrightarrow$	
$X_{IFD} = E(K_{IFD}, RND_{IFD} \parallel RND_{ICC})$	$\longleftarrow$	Response [RND_{ICC}]
MUTUAL AUTHENTICATE [X_{IFD}]	$\longrightarrow$	$RND'_{IFD} \parallel RND'_{ICC} = D(K_{ICC}, X_{IFD})$
		IF ($RND_{ICC} = RND'_{ICC}$)
		THEN terminal authenticated
		ELSE terminal not authenticated
		$X_{ICC} = E(K_{ICC}, RND_{ICC} \parallel RND'_{IFD})$
$RND'_{ICC} \parallel RND'_{IFD} = D(K_{IFD}, X_{ICC})$	$\longleftarrow$	Response [X_{ICC}]
IF ($RND_{IFD} = RND'_{IFD}$)		
THEN smart card authenticated		
ELSE smart card not authenticated		
T = tag of the required access data		
GET DATA[T]	$\longrightarrow$	Z = fetch the access data referenced
		by the tag
Z = access data	$\longrightarrow$	Response [Z]

Sequence Chart 9.2   Sequence of events for mutual authentication of the access terminal and smart card using a challenge–response process, followed by a request for the access data. This is the usual sequence in operational use

otherwise a suitable return code is sent back. The terminal evaluates the information received in this manner and grants access to the room if the result of the comparison is positive.

The security of this sequence can be defeated by an attacker equipped with suitable technical equipment. To do so, the attacker must interrupt communication with the smart card between the successful mutual authentication and the transfer of the access data, without this interruption being noticed. After this, the attacker sends access data suited to his purpose back to the terminal in real time in response to the request from the terminal. If this data has the correct syntax and the proper access code, the terminal will enable access.

The access data could be protected by a cryptographic checksum, but this would not provide a particularly high level of security. This is because the protected data could have been spied out of a genuine smart card after successful authentication with a genuine terminal and then played back using the sort of manipulation of the data transmission process described above. This would effectively bypass the protection provided by the cryptographic checksum.

In any case, the smart card application should be structured somewhat differently for cards used for controlled access to high-security areas. A good approach would be to weave the access data into the mutual authentication process. This would automatically protect the access data against playback, eavesdropping and skimming. Sequence Chart 9.3 shows the structure of the modified authentication data and a command sequence that could be used in this case. A similar method is used in the European standard for electronic purse systems (EN 1546).

Terminal (IFD)		Smart Card (ICC)
SELECT [AID]	$\longrightarrow$	
	$\longleftarrow$	Response

Terminal (IFD)		Smart Card (ICC)
T = tag of the required access data		
GET CHALLENGE	$\longrightarrow$	
$X_{IFD} =$	$\longleftarrow$	Response [RND_{ICC}]
$E(K_{IFD}, RND_{IFD} \parallel RND_{ICC} \parallel T)$		
MUTUAL AUTHENTICATE [X_{IFD}]	$\longrightarrow$	$RND'_{IFD} \parallel RND'_{ICC} \parallel T =$
		$D(K_{ICC}, X_{IFD})$
		IF ($RND_{ICC} = RND'_{ICC}$)
		THEN terminal authenticated
		ELSE terminal not authenticated
		Z = fetch the access data referenced
		by tag T
		$X_{ICC} =$
		$E(K_{ICC}, RND_{ICC} \parallel RND'_{IFD} \parallel Z)$
$RND'_{ICC} \parallel RND'_{IFD} \parallel Z =$	$\longleftarrow$	Response [X_{ICC}]
$D(K_{IFD}, X_{ICC})$		
IF ($RND_{IFD} = RND'_{IFD}$)		
THEN smart card authenticated		
ELSE smart card not authenticated		
Z = authentic access data		

Sequence Chart 9.3   Sequence of events for mutual authentication of the access terminal and smart card using a challenge–response process with embedded access data query. The advantage of this form of communication is that the information exchange process used for the access data is cryptographically protected

However, this approach has a distinct drawback. The original solution can be implemented using standard smart card operating system commands for file-based applications, but the commands needed for the improved solution using authentication commands with expanded functionality must be specifically developed for the smart cards used in the system. It presupposes that the smart card operating system supports the executable program code necessary for this purpose. However, it does not require a particularly large amount of development effort, so it can certainly be regarded as an acceptable approach.

The command sequences used during the administrative phase are independent of the specific configuration of the processes used in the operational phase. Personalization must be performed during the administrative phase before the smart cards are issued to the users. This involves writing the smart card identification number, the secret authentication key, and the access data for the individual user to each smart card. It is assumed that personalization is performed in a secure environment, so the data can be transferred to the smart card as plain text without compromising security.

Authentication of the outside world by the smart card is sufficient for attaining the security state necessary for writing data to the smart card. This is done using the EXTERNAL AUTHENTICATE command as shown in Sequence Chart 9.4. For security reasons, the

Terminal (IFD)		Smart Card (ICC)
SELECT [AID]	$\longrightarrow$	
	$\longleftarrow$	Response
GET CHALLENGE	$\longrightarrow$	
$X = E(K_{ICC}, RND)$	$\longleftarrow$	Response [RND]
EXTERNAL AUTHENTICATE [X]	$\longrightarrow$	$X' = E(K_{ICC}, RND)$
		IF $(X = X')$
		THEN terminal authenticated
		ELSE terminal not authenticated
	$\longleftarrow$	Response
Tag T of the access data Z		
PUT DATA[T ‖ Z]	$\longrightarrow$	Store the access data Z identified by T
	$\longleftarrow$	Response

Sequence Chart 9.4 Sequence of commands between an administration terminal and a smart card for unilateral authentication using a challenge–response process to obtain write access privileges for all data objects and subsequently transfer an access data set

key used for this purpose is not the same as the key used for mutual authentication in the operational phase. After unilateral authentication, all necessary data can be entered sequentially in the card using PUT DATA with the tags as parameters. Information regarding the smart cards that have been loaded in this manner and the associated data is then transferred in real time to the background system. These data sets are generally referred to as 'response data'.

An administration workstation is necessary if the access privileges of a smart card must be modified at some point after the smart card has been issued. This workstation is operated by an authorized person. Following successful authentication, the terminal can be used to modify the data in the smart card as appropriate. To make the process of assigning access privileges easy for the operator, the computer has a special user interface that lets the operator work with terms such as 'Archive entrance door' instead of TLV-coded data objects. The computer controls the administration terminal. The process necessary for this step corresponds to the personalization process with the restriction that the only function available here is editing the access data. Ideally, the database of the background system should be updated immediately during this administrative process in order to ensure that the access privilege information stored in the background system is always current.

The application described here is normally one of several applications in a company ID card, which can also include other applications such as encryption and decryption, a digital signature, single sign-on, and personal identification. It has a relatively simple structure, but it is nevertheless flexible and able to withstand simple attacks. For instance, it is resistant to spying out the data because access to the data requires at least successful mutual authentication of access terminal and the smart card. Some other typical uses for this application include car park access cards and admission control for a wide variety of events.

9.3   Telemetry Module

A while ago, I received an e-mail message from a Spanish company that had developed a small telemetry module with an integrated smart card for security functions. Starting from this initial contact, a brisk exchange of information developed over the course of an extended length of time, in which both sides obtained new insights into the field of secure public data communication. I would like to present these insights here.

The telemetry module in question has a user-programmable microcontroller and a low-power transmitter with a range of a few kilometres. With optional accessory components, the data can also be sent in the form of short message service (SMS) messages via a GSM transmitter module. The main advantage of this telemetry module is that data transmission is cryptographically secured to prevent eavesdropping and manipulation. The data security functionality is provided by a smart card in ID-000 (Plug-In) format,[1] which can be removed from the telemetry module if necessary.

The module is intended to be used for secure unidirectional transmission of small amounts of various sorts of data to a central location. A typical application would be transmitting flue gas measurements directly from an industrial chimney to an environmental data centre. Another potential application is automatic acquisition and transmission of data regarding utilization of public car parks to a traffic information system. Both of these applications involve a certain amount of preparation of the acquired data by the sensor stage and transmitting the prepared data in a manner that is secure against manipulation.

The data to be transmitted must be split into blocks according to the cryptographic algorithm used, and if necessary it must be padded to a integral number of blocks. Encryption is performed using the AES cryptographic algorithm in CBC mode with a 128-bit key. This configuration is sufficient to provide adequate protection against attacks for many years to come. Figure 9.2 shows the basic architecture of the original secure communication mechanism.

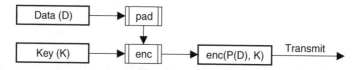

Figure 9.2   Block diagram of the original security mechanism of the telemetry module described in the text. The padding function ('pad') pads the data to an integer number of blocks for the cryptographic algorithm so the data can be encrypted in CBC mode by the encryption algorithm in the 'enc' block. The encrypted data is then transmitted on a radio channel to a central recipient

When the telemetry modules were first put into operational service, it was found that the original straightforward approach to secure communication gave rise to a host of problems. One of the critical difficulties was that it was impossible to be sure that the data was correct after decryption, since the encrypted data could consist of arbitrary byte

---
[1] See Section 1.2

sequences. Consequently, the data recipient was forced to perform tests at the application level to ensure that the received and decrypted data matched the original data. In an ideal system, this should be ensured reliably by the transmission layers.

There were also several other problems. First, the transmitter did not add a sender ID to the transmitted messages, so it was difficult for the receiver to process messages from multiple transmitters using individual encryption. The only way to perform decryption was to use trial and error, which is a rather makeshift solution. Another related shortcoming was that the system supported only one key per telemetry module.

In typical applications, the data is usually not truly sensitive in a security sense. However, a serious weakness of the original version of the system was that it did not provide any protection against replay attacks or corrupted transmissions. This sort of attack is particularly easy with radio data links, and it does not require very much effort or expense.

With the exception of vulnerability to replay attacks, these shortcomings (no matter how significant in an objective sense) would not have an especially severe impact on individual applications tailored to specific uses. However, the telemetry module could not provide reliable, secure data transmission as an intrinsic capability for any desired application without some fundamental changes to its design. Figure 9.3 shows the result of the improvements.

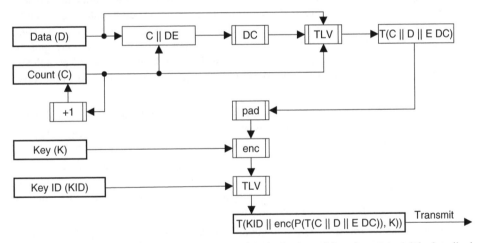

Figure 9.3   Block diagram of the improved security mechanism of the telemetry module described in the text. The 'EDC' function calculates an error detection code, the 'TLV' function structures the data received by the module, the 'pad' function pads the data to an integral number of blocks for the cryptographic algorithm, and the 'enc' function encrypts the data in CBC mode

A counter incremented after each data transmission was added to defend against replay attacks. If the receiver evaluates the counter value, it can detect duplicated or missing data packets resulting from replaying previous transmissions. This provides reliable defence against replay attacks. The counter value is placed at the beginning of the user data so the encrypted data packets will be different even if the user data is the same. All data objects are TLV coded so their length can be modified or additional data objects can be added without requiring changes to be made to the evaluation routine at the receiver end.

It is not necessary to include the identification number of the telemetry module, since the identification number of the key used for encryption, which is also included, is usually sufficient. If the telemetry modules are assigned individual keys, they can be identified uniquely by their key IDs. Using key IDs also makes it possible to place several keys in each telemetry module. If multiple keys are present, they can also be swapped periodically (among other options).

The process for receiving an encrypted message operates reverses the sequence of the process for transmitting the message. It first uses the TLV codes to extract the key ID and encrypted data block from the message. After this, it decrypts the data block using the key identified by the key ID and the associated cryptographic algorithm. The individual data items in the decrypted data block can then be extracted using the TLV coding. If the required TLV data objects in the data block cannot be recognized, this is a sign that the decryption was not correct. This should cause all further operations on this data block to be cancelled. If all of the necessary data objects can be recognized correctly, the next step is to check whether the error correction code can be calculated correctly. If it can, the decryption is correct.

Next, the receiver checks whether the counter value is plausible. A suitable control mechanism is to accept only messages with counter values greater than the counter value of the last message received. This ensures that no duplicated counter value can be interpreted as being valid. Depending on the required level of security, the counter values can also be checked for strict adherence to an increasing sequence without any gaps. If the counter check is completed successfully, the data has been transferred completely and without any errors, and it can be passed to the ultimate application in the receiver system.

The cost of the necessary improvements to the smart card lay within reasonable bounds relative to the considerable improvement of security and reliability. This particular application was based on Java smart cards, so the changes could be made quite quickly.

The fact that the smart cards can be removed from the telemetry modules makes administration of the smart card applications considerably simpler. The data in the cards can be configured for a particular use by inserting them in a PC terminal and using standard smart card commands to modify the data stored in the cards.

If it is necessary to transmit data immediately after a particular event without thereby revealing that some event has occurred, this can be realized using a simple mechanism. It consists of having the telemetry module transmit empty messages at irregular intervals in addition to actual event data. This makes it rather difficult for an outside observer to recognize which specific events lead to transmission of encrypted messages. Another significant consideration here is that the length of the transmitted messages does not depend on whether they contain real information or are only dummies.

9.4 Business Card

A friend who works in the telecommunications business recently showed me a smart card that he received as a combined admission pass and business card when he attended a trade fair. He proudly told me that he only had to hold the card next to a terminal to gain admission to the fair, which obviously meant it was a contactless smart card. The card could also be used to quickly convey his name, postal address and e-mail address to attendants at exhibitor stands. My friend also reported that contactless communication was only real innovation in the pass used at this year's fair, since contact smart cards with comparable functions were issued to visitors to the previous year's edition of the fair.

Visitors to the fair had to register in advance and provide their personal data in order to obtain the cards. This could be done online or by using terminals in the entrance area of the fair. Before entering the fair, each registered visitor received a contactless smart card printed with the visitor's name and fitted with a neck strap. After being admitted to the fair, visitors could wander from one stand to the next, and if they wished they could use portable terminals issued to the stand attendants to transfer their personal data to the attendants. As usual, visitors received suitable promotional material after the fair, along with many friendly telephone calls to enquire when they expected to convert their purchasing intentions into actual orders.

The major benefit of this card is that exhibitors do not have to enter data into their databases from the printed business cards of stand visitors. This represents a considerable time savings. The fair visitors also benefit from the fact that they do not have to carry around stacks of printed business cards, which are usually thrown away immediately after the data has been recorded.

I found this smart card application interesting, so I started analysing the card to learn more about it. This was not particularly difficult. It proved to be a memory card with 256 bytes of EEPROM and a data transmission protocol compliant with ISO/IEC 14 443 Type A. Surprisingly enough, the data was not protected and could be read and written freely. All the data stored in the card could be read using simple equipment. The content of the memory is portrayed in Table 9.4.

Table 9.4 An excerpt of the memory content of a contactless smart card issued as an electronic business card for visitors to a trade fair. In addition to the data shown here, the card contains the visitor's mobile telephone number and his or her position in the company, postal address, date of birth, and registration number

	x0	x1	x2	x3	x4	x5	x6	x7	x8	x9
1y	...	...	...	...	...	...	...	...	...	...
2y	'80'	M	r	.	'81'	V	a	l	e	n
3y	t	i	n	'83'	S	t	e	i	n	b
4y	e	r	g	'88'	S	t	e	i	n	b
4y	e	r	g		T	e	c	h	n	o
6y	l	o	g	i	e	'89'	+	4	9	8
7y	9	1	2	3	1	6	2	0	'8A'	V
7y	a	1	e	n	t	i	n	.	S	t
7y	e	i	n	b	e	r	g	@	g	m
7y	x	.	n	e	t	...	...	...	...	...
8y	...	...	...	...	...	...	...	...	...	...

The sheer volume of data brought to light by this analysis astonished me. Some of the information was not really necessary for business contacts in the context of a visit to a trade fair. Although this data had been given more or less voluntarily by the visitor prior to the fair, it was rather surprising that the full set of data was recorded in the smart card.

There are two use cases for the data stored in the card: admission control and the electronic business card. For admission control, the attendants at the entrance doors might need the visitor's date of birth as well as the card, depending on the situation, in order to have at least some idea of whether the card had been improperly given to another person. However, the visitor's date of birth is irrelevant for visits to the actual stands at the fair. For this reason, the data originally obtained from the visitor should be tied to a specific purpose, as is also required by commonly applicable data protection regulations.[1]

Another issue that arises with this card is expression of intent, which affects the entire realm of contactless smart card technology. As no specific condition (such as a PIN verification) must be fulfilled to read data from the card, anyone equipped with a suitable terminal can read the contents of the card without the permission of the user. If someone were to do this systematically and unobtrusively at the fair, he could obtain a considerable collection of personal data. In all the hustle and bustle of the fair, visitors would probably not notice that their cards were being read.

The same problem arises in similar situations with passports containing data storage media that can be read using wireless technology. Here as well, sensitive personal data (such as fingerprints) can be read surreptitiously at short distances if suitable protective measures are not taken. For this reason, methods have been developed to prevent unobserved reading of electronic passports.

The method used for this purpose, which is called *Basic Access Control*, is based on using data that can only be read optically from the passport as a sort of key. This key forms the basis for mutual authentication of the terminal and the smart card. The terminal is not granted access to the passport data via the wireless link until this authentication has been completed successfully.

This intentionally introduced media discontinuity makes it impossible for the passport to be read surreptitiously via the wireless interface, since the key must be known before this can be done and this requires optically readable information from inside the passport.

This principle can also be adapted for use with contactless business cards, although in this case the objective is not to construct a highly secure application, but instead to provide relatively weak protection against unauthorized reading of the card data. Sequence Chart 9.5 shows a possible variant of Basic Access Control for contactless business cards.

One option would be to use the name printed on the card as a key for authenticating the terminal relative to the smart card. For this purpose, the name would have to be entered manually in the terminal or read using an (OCR) device. It could then be converted into a suitable key format using a simple mapping process. In response to a request from the terminal via the wireless interface, the smart card would send a random number to the terminal, which would then encrypt it using the key and return the result to the smart card. The smart card would know the name and thus the key, so it could duplicate the encryption performed by the terminal. If the two cryptographic results matched, the

---

[1] See Section 4.1

IFD (Terminal)		ICC (Smart Card)
Read optical card data and use it to calculate K_{IFD}		
GET CHALLENGE	$\longrightarrow$	Generate RND_{ICC}
Generate RND_{IFD}	$\longleftarrow$	Response $[RND_{ICC}]$
$X_{IFD} = E(K_{IFD}, RND_{IFD} \parallel RND_{ICC})$		
MUTUAL AUTHENTICATE $[X_{IFD}]$	$\longrightarrow$	$RND'_{IFD} \parallel RND'_{ICC} = D(K_{ICC}, X_{IFD})$
		IF $(RND_{ICC} = RND'_{ICC})$
		THEN terminal authenticated; grant read access to the EF containing the business card data
		ELSE terminal not authenticated; block read access to the EF containing the business card data and terminate
		$X_{ICC} = E(K_{ICC}, RND_{ICC} \parallel RND'_{IFD})$
$RND'_{ICC} \parallel RND'_{IFD} = D(K_{IFD}, X_{ICC})$	$\longleftarrow$	Response $[X_{ICC}]$
IF $(RND_{IFD} = RND'_{IFD})$		
THEN smart card authenticated		
ELSE smart card not authenticated		

Sequence Chart 9.5 Procedure for mutual authentication of a terminal and a smart card. The authentication key of the terminal is computed exclusively from optically readable data printed on the body of the card. This scheme is based on the operating principle of Basic Access Control for e-passports with contactless chips

smart card would know that the terminal also knew the name printed on the card and it would grant access to the business card data. The drawback of this procedure is that it requires an extensive sequence of commands between the terminal and the smart card. It also means that the smart cards must be processor cards, which are more expensive than memory cards.

Another option would be to print a visually readable key on the back of the smart card. Such a key could consist of six to eight characters or digits so it could be read and typed in easily. Another alternative worth considering is storing the visually readable key in the form of a bar code, which can be read reliably with suitable equipment. Such a system would not withstand a brute-force attack on the key, since the key space is too small. However, this should be quite acceptable as applications of this sort do not need a high level of security.

Regardless of what form this key takes, it is located on the back of the card so it cannot be read surreptitiously. The card must be turned around to read the key, and the card holder will only do this consciously. The business card data stored in the smart card is encrypted using the key. In this scenario, all the data stored in the card can be read by the terminal without any restrictions, but the data cannot be used without the key.

The use case would thus take the following form. If a visitor wishes to leave his personal data at a stand, he holds his smart card next to a terminal so the terminal can read the business card data. Next, the visitor gives the key printed on the back of the card to a stand attendant so the data read from the card can be decrypted. This general features of this scenario are show in Sequence Chart 9.6.

IFD (Terminal)		ICC (Smart Card)
Command [read c]	$\longrightarrow$	Fetch c from memory using c = E(v, k)
Enter key k$'$	$\longleftarrow$	Response [c]
Transform k$'$ into k		
v = D(c, k)		
Further use of business card data v		

Sequence Chart 9.6 Use case process for transferring business card data (v) via a wireless interface to a terminal at a fair stand. The encrypted business card data is designated c. The key (k) printed on the card is conveyed to the terminal by manual entry or OCR scanning

This solution prevents unauthorized use of the data, and it has yet another advantage: the data stored in the card can be partitioned for specific uses. A data set specifically prepared for the 'admission control' use case can be stored in the card for use by the entrance attendants. This data can then be encrypted using a key known only to the admission control terminals. This way all data necessary for admission control can be checked quickly and reliably. A second data set is can also be stored in the card, consisting of the actual business card data encrypted using the key printed on the back of the card. This data can then be used by the stand attendants. This creates a clear division between the two use cases, and the individual entities receive only the data that is essential for their specific purposes. This solution could also be integrated into the existing system without major modifications.

9.5 Theft Protection Card

For more than two decades, supplementary hardware has been used to protect software against unlawful copying. This typically takes the form of a dongle that is physically and logically connected to the computer that hosts the protected software. The connection is usually made using a standard interface, such as RS232 or USB. As using dongles repeatedly leads to problems for a wide variety of reasons, and because they entail additional costs, there has been a growing trend in recent years to use software registration instead of dongles. This is made easier by the fact that a many computers these days are linked to the Internet.

However, there are many use cases in which a dongle is certainly worthwhile. These cases are also fully independent of the social and political debate regarding the pros and cons of copy protection, and they yield clearly visible benefits from the perspective of the end user. A good example of this is using a smart cards to provide theft protection for a car radio. The card is inserted into a simple card reader in the radio, and the driver takes the card with him after parking the car. The security is provided by the fact that the radio cannot be used unless the smart card is inserted. A stolen radio without the smart card is thus worthless for a thief, which sooner or later leads to such radios being purloined less often. The details of how this form of theft protection for car radios actually works are not public knowledge, but it is reasonable to assume that it is modelled on the long-established copy protection mechanism of PC dongles, since it provides a solution to a similar sort of problem.

From the description of 'The Smart Card Simulator' (TSCS)[1], it can be seen that a paid version with copy protection was planned in addition to the free version. As I came to know and appreciate the idea of public licence software shortly after the first publication of TSCS, it is now available under a GNU Public License (GPL) instead, and my ideas regarding copy protection have been permanently shelved. Nevertheless, it is interesting to have a look at the ideas from that time, since they are still valid and can be put to good use.

Software protection using additional hardware always operates on the principle of storing information or portions of the program necessary for running the software in a secure form in some sort of hardware. To minimize the level of abstraction, this can be described in more detail using a specific example. Here we assume that there is a device that requires protection against theft, and this protection is implemented using a smart card. The basic idea is that the device cannot be used if the smart card is not inserted in the card reader.

The simplest approach is to have the device read a 'magic number' from the smart card and compare it with a stored number. If the two numbers match, the device will be enabled for use. This is illustrated graphically in Sequence Chart 9.7. Naturally, each device must have a different number, and the number must be long enough to prevent it from being guessed. A length of 8 bytes gives 2^{64} possible numbers, which is enough to defeat any form of brute-force attack if we assume a transaction time of at least 1 second for each attempt.

Device with Terminal (IFD)		**Smart Card (ICC)**
. . .		
Fetch number M_{IFD} from memory		
Command [fetch number M_{ICC}]	$\longrightarrow$	Fetch number M_{ICC} from memory
IF ($M_{IFD} = M_{ICC}$)	$\longleftarrow$	Response [M_{ICC}]
THEN execute rest of program		
ELSE terminate program		
. . .		

Sequence Chart 9.7 Querying a smart card for a magic number M as a simple form of theft protection

Another approach is to add challenge–response authentication of the device and smart card, similar to the process described elsewhere in this book.[2] However, security measures of this sort have the disadvantage that they can be bypassed by manipulating the software in the device. If the challenge occurs only once, it is sufficient to patch out the challenge routine in the device software with a series of assembly-language instructions that do nothing, such as NOP (no operation). Consequently, a more sophisticated

[1] See Rankl and Effing (2002)
[2] See Section 9.1 and Section 9.2

approach is to scatter numerous challenge routines throughout the program code of the device software.

Another form of protection is to relocate essential parts of the device software to the smart card. For instance, a table of branch targets can be relocated to the smart card. In this case, the software in the device will run properly until it reaches a certain point where it must branch to a different location to continue execution. The branch targets are stored in the smart card and read from the card by a special function. The software stored in the device is thus incomplete, so the branch instructions cannot be 'NOPped out' with a debugger. This arrangement is portrayed in Sequence Chart 9.8.

Device with Terminal (IFD)		Smart Card (ICC)
...		
Instruction n in program code		
Command [fetch branch target i]	$\longrightarrow$	Fetch address A for branch target i
Branch to address A_i	$\longleftarrow$	Response [A_i]
...		

Sequence Chart 9.8 Theft protection based on placing a table of branch targets and target addresses A_i in a smart card

Yet another option is to relocate entire functions to the smart card instead of just a table of branch targets. If the result of one of these functions is needed, the necessary input data is sent to the smart card, which in turn returns the computed result to the device. The best way to use this approach is to relocate complicated algorithms to the smart card, since they are more difficult to reproduce by manipulating the software in the device. This option is illustrated in Sequence Chart 9.9.

Device with Terminal (IFD)		Smart Card (ICC)
...		
Instruction n in program code		
Input data IN of function F		
Command [IN]	$\longrightarrow$	OUT = F(IN)
Use OUT in ongoing program execution	$\longleftarrow$	Response [OUT]
...		

Sequence Chart 9.9 Relocating calculations to the smart card as a way to bond a device and a smart card to provide theft protection

Relocating functions is often used as a form of copy protection for software in mobile telephones. In this case, the Java applet in the telephone delegates important calculations to the SIM. The methods stored in the smart card are called using remote method invocation (RMI) as specified in the Security and Trust Services API (SATSA) in accordance with JSR 1227. To make it more difficult to analyze the Java byte code stored in the mobile telephone, an obfuscator can be used to camouflage the functionality of the finished code. An obfuscastor rearranges the code so it is difficult to understand after being disassembled. This does not provide any real protection against hard attacks, but it does increase the cost of the analysis.

A somewhat different approach to theft protection is to encrypt segments of the software to be protected. When it is necessary to execute an encrypted segment of the software, it is sent to the smart card where it is decrypted using a suitable cryptographic algorithm and a secret key, after which the decrypted code is returned to the program that issued the request. There the decrypted program segment is run and then deleted from memory immediately after it has been used. A device-specific key must be used for this purpose, as otherwise any desired smart card could be used to decrypt the software for the application. This principle is shown in Sequence Chart 9.10.

Device with Terminal (IFD)		**Smart Card (ICC)**
. . .		
Instruction n in program code		
Fetch eP_i = enc(PrgCode$_i$, K) from		
memory		
Command [eP_i]	$\longrightarrow$	P_i = dec(eP_i, K)
Execute program code P_i	$\longleftarrow$	Response [P_i]
. . .		

Sequence Chart 9.10 Using a smart card to decrypt segments of the program code before it is executed in the device. This is a very flexible and general-purpose form of theft protection. If a device-specific key is used, the smart card and the device are tied to each other and cannot be interchanged with other devices or cards

In practice, the methods described here are often combined and incorporated into the program code of the device in several locations. In addition, the code is intentionally designed so some queries occur frequently while others occur only rarely, in order to increase the amount of time necessary to analyse communications while still ensuring that the absence of the smart card can be recognized quickly.

Efforts are also made to hinder attacks based on manipulating the software in the device. This can involve using established mechanisms such as checksums and signatures calculated from segments of the program code, as well as analyzsing program execution time. The objective here is to increase the cost of an attack to the point that it is no longer sufficiently attractive for a potential attacker. If this can be achieved, the protection is effective. An important aspect here is that theft protection systems, and incidentally copy protection systems as well, must always be constructed such that a successful attack on an individual device cannot lead to discrediting all the devices, but instead remains limited to breaching the security of the attacked device.

9.6 Admission Pass

Some time ago, a horticulture exhibition lasting all summer was organized at considerable expense in a large city in southern Germany. The exhibition will probably go down in the annals of horticulture as a colossal waste of effort, but the season passes issued for the exhibition are nevertheless interesting from a technical perspective.

The season passes were contactless cards in ID-1 format with a barcode instead of a chip. This was a 2-of-5 interleaved barcode, which is a purely numeric code with parity and an optional check digit at the end, which was not used in this case. A colour photo of the pass holder was printed on each card using thermal transfer printing. The season passes could be ordered via the Internet or at the ticket counter and then be picked up a few days later. Figure 9.4 shows a sketch of the front of the pass.

Figure 9.4 Sketch of the front of a season pass for the exhibition mentioned in the text. The back of the pass was fully occupied by an event logo without any additional information. The number printed at the bottom left is the numerical representation of the barcode

The use scenario was rather simple. Attendants armed with morose expressions and hand-held terminals fitted with barcode readers were stationed at the entrances. They checked the resemblance between each potential visitor and the photo on the pass while using the terminal to scan the barcode printed on the pass. If the photo was a reasonably good match to the person, they then checked when the person had last visited the exhibition. If the interval was a few days, the person was granted admission, and otherwise the person was once again checked quite carefully against the photo. Visitors could leave the exhibition without being registered by revolving doors.

The admission charge was more than 10 euros, so a fraud scheme developed rather quickly. A genuine pass holder passed through the admission control to the exhibition grounds and then handed his or her pass to a friend waiting outside the chain-link fence. The friend then entered the grounds, taking care to use a different entrance or at least avoid passing by the same entrance attendant.

This fraud vulnerability was discovered fairly quickly, and as a countermeasure the hand-held terminals were linked online to the background system so the attendants could receive information about the most recent entry time of the pass holder after reading the barcode. The response of the swindlers to this measure was to have the false pass holder enter first, followed by the genuine pass holder. As the genuine pass holders could prove their identity with personal ID cards if necessary, the attendants ultimately had no other choice than to admit them to the exhibition. After all, it was always possible that a visitor had left the exhibition grounds briefly right after entering. One of the standard answers visitors gave in response to questioning by an attendant was that they had forgotten their camera in the car.

As the total extent of the fraud remained within reasonable limits, the organizers continued to use the admission pass system according to the original concept without any modifications.[1]

The real risk with barcode cards is not the sort of minor cheating described above, but instead systematic large-scale forgery. This does not mean cloning existing cards, but instead generating new cards with valid serial numbers that were never produced officially. In the above case, all that would be necessary to produce clones is a thermal transfer printer for the cards. If a sufficient number of blank cards are used as a basis for the cloning operation, it can take some time before this sort of attack is discovered.[2]

The chance of success of this form of attack depends on several factors. One of them is whether all officially issued cards are registered in the background system, accompanied by recording the serial numbers of all valid cards in a 'whitelist' that is constantly updated and synchronized with the terminals of the entrance attendants. This reliably prevents unauthorized production of new 'valid' serial numbers. However, this solution is only viable if the cards and their serial number are recorded in the background system in real time when they are produced, and the copies of the whitelist in the terminals must be synchronized with the master list at least once a day.

If the cards are produced in more than one location, this is difficult to implement. The quality of the printing on the cards issued for the landscaping exhibition was low and varied considerably among the cards, which suggests that they were produced at several locations using inexpensive desktop printers and prepared blank cards. In this sort of scenario, serial numbers are typically generated at each location where the cards are produced and then printed on the cards.

An analysis of the serial numbers on the cards showed that the leading digits contained a reference to the individual card producers. This strongly suggests that each of the producers was assigned a block of numbers and was allowed to independently produce cards with individual serial numbers in this block.[3]

If it is not possible to report the produced cards to the background system, another option is to use a check digit generated using an algorithm with a secret key. This can prevent unauthorized generation of valid cards, since the secret key is not known to potential attackers.

In any case, a more secure solution would be to use contactless smart cards instead of technically unsophisticated barcode cards. Contactless smart cards also impose fewer restrictions on the graphic design of the card, since the chip and antenna inside the card replace the barcode printed on the card. They also provide better protection against forgery, since the technology is much less widely distributed than printing barcodes on plastic cards.

---

[1] Of course, scoffers claimed that the exhibition was so dull that most visitors had no interest in seeing it more than once or twice

[2] See Section 7.3

[3] See Section 7.1

Of course, it does not help much to use contactless smart cards if users can simply pass them through a fence to other potential visitors after they enter the exhibition grounds. As a countermeasure in this case, it would be advisable to also record the visitors leaving the exhibition. This is a quite reliable way to prevent several people from using a single card to gain admission to the exhibition grounds one after the other.

It would also be possible to use biometric features, such as fingerprints or facial recognition, to uniquely associate individual cards with specific persons. However, using biometric techniques to identify individual persons for applications such as this is a highly controversial social and political issue.

9.7 PKI Card

A company ID card with a digital signature function can provide useful benefits for companies spread over several sites and in situations involving working with suppliers. Such a card need not necessarily be fully compliant with digital signature standards and regulations. In many cases, it is sufficient to add a digital signature function to existing smart cards used as company ID cards. This presupposes that sufficient memory space is available and that the smart card operating system supports asymmetric cryptographic algorithms. However, both of these requirements can generally be satisfied without any technical gymnastics.

One of my former neighbours works for a company that upgraded from chipless plastic ID cards to a system based on smart cards. With both types of cards, a photo of the employee is printed on the front of the card using thermal transfer printing. In the course of our regular Saturday-morning strolls to the bakery, my neighbour told me about interesting episodes related to the new company ID cards, extending from the initial concept to the first months of operational use.

The essential reason for introducing the new system was the desire to establish fraud-proof internal communications. This was also supposed to include electronic work processes such as requests for leave, flexitime corrections, requests for travel on company business, and internal orders. A parallel objective was to increase the level of security of communications with suppliers. Here it can be remarked that these improvements were integrated quite successfully into the existing systems of the company.

Particularly in the case of signature cards, the relevant standards (such as ISO/IEC 7816-4, ISO/IEC 7816-8 and CWA 14 890) provide a broad range of options. This challenge was mastered in this particular case by taking a quite rigorous approach to restricting the application to the essential functions of a signature card, which are generating and verifying signatures. For this reason, no externally configurable options were provided and the signature key and associated public key were rigidly defined by the application.

Authentication of the terminal,[1] which is often performed before PIN verification, was also omitted. Key storage was based on the provisions of CWA 14 890.

These simplifications make it possible to fashion the offcard portion of the smart card commands necessary for the application in a form that is significantly easier to comprehend than is usually the case. Sequence Chart 9.11 and Sequence Chart 9.12 show

---

[1] See Section 5.10

the essential aspects of signature generation and signature verification in a simplified form.

IFD (Terminal)		ICC (Smart Card)
SELECT [AID]	$\longrightarrow$	Select PKI application
	$\longleftarrow$	Response
PIN entry by user		
VERIFY [PIN]	$\longrightarrow$	Compare the PIN passed with the command with the stored PIN
	$\longleftarrow$	Response
Compute hash value h = H(m)		
PSO: COMPUTE DIGITAL SIGNATURE [h]	$\longrightarrow$	s = S(h, sk)
	$\longleftarrow$	Response

Sequence Chart 9.11 A typical process for generating a signature from a hash value in a signature application. PERFORM SECURITY OPERATION (PSO) is used in the COMPUTE DIGITAL SIGNATURE function. The hash value h calculated from the original message m is sent to the smart card and signature generation is invoked in the smart card. The signature s is contained in response to the command

IFD (Terminal)		ICC (Smart Card)
SELECT [AID]	$\longrightarrow$	Select PKI application
	$\longleftarrow$	Response
PSO: SELECT [h]	$\longrightarrow$	
	$\longleftarrow$	Response
PSO: VERIFY DIGITAL SIGNATURE [s]	$\longrightarrow$	r = V(h, s, pk)
IF r = ok	$\longleftarrow$	Response [r]
THEN signature correct		
ELSE signature not correct		

Sequence Chart 9.12 A typical process for verifying a signature s in a signature application. PERFORM SECURITY OPERATION (PSO) is used in the HASH function to transfer the hash value h to the smart card and in the VERIFY SIGNATURE function to verify the signature. The public key pk is selected implicitly by the application. The response to the PSO command contains the result r of the verification

However, the most striking feature of this signature application lay in a different area. The signature function of the smart card could be used with a variety of PC applications, and as usual in this environment, before actually signing a document with a digital signature the user had to enter a PIN code as a clear expression of his or her intent to generate a signature. Unfortunately, there was no uniform user interface for the PIN input function. Depending on the specific PC application, the user might be asked to enter a PIN, enter a password, enter a secret number, or enter a personal identification number. As a result, users occasionally entered an incorrect password – usually one of the many passwords modern PC users have to remember in their day-to-day work – instead of the correct PIN for the signature function.

After three failed attempts to enter the correct PIN code, the error counter reached its maximum value and the smart card was blocked. This affected not only the signature function, but also all other functions of the card. This resulted from the fact that only one PIN was assigned to each card, which is actually a good idea in theory. The smart cards had a suitable deblocking function that could be used to store a new PIN code in the card and reset the error counter reset to zero. The user interface of the PC application for this purpose was perfectly clear, so the affected smart cards could be unblocked by their users without any difficulty. However, general acceptance of the application suffered noticeably from the lack of a clearly defined user interface for PIN entry.

There was also a second faux pas in this smart card application. The PIN and PUK codes were set to the same value during personalization, presumably for the sake of simplicity. An unavoidable consequence of this was that the card was permanently blocked if its user forgot the PIN code. Unfortunately, there was no provision for unblocking cards via the background system, so forgetting the PIN code meant the kiss of death for the smart card, which thus had to be replaced by a new one. This entailed costs for issuing a new card for each blocked card, which could have been avoided if the system had been designed somewhat more intelligently.

Several lessons can be deduced from this story. The first lesson is that a clear, easily understood user interface is an essential quality feature of a good offcard applica- tion. If the user interface is not clear and consistent – for example if several dif- ferent terms are used for the same thing – the result is user uncertainty, which in turn leads to entry errors. As a result, the system operator will be confronted with significantly higher operating costs than were originally planned. A field trial with a limited number of users is a suitable means to discover potential problems at an early stage. However, the responsible parties must have the courage to factor the cost of changes that can be expected as a result of the field trial into the overall project plan.

The second lesson that can be deduced from the above example relates to the necessary administrative functions. With closed applications such as company ID cards, it is useful to be able to access the functions in the smart cards, including the functions related to security, from the background system in certain situations. In the case of the company ID card described here, it would certainly be possible to ascertain the identity of any employee whose card has been blocked owing to too many false PIN code entries and allow the employee to enter a new PIN code using a special administration terminal. This would avoid the need to issue a new card when the card holder has forgotten the PIN code.

9.8 SIM Card

When hunters feed game animals in the forest, they need to know how many animals (such as deer) visit the feeding station. Ideally, they want to have this information in near real time. They would also like to know whether the feeding station draws a lot of animals and when the feed needs to be replenished. There are a few small manufacturers that supply equipment suitable for registering game animals at feeding stations. However, this interesting problem also encourages people to develop their own solutions, and as a result I came into contact with a person who had built a 'game animal registration device'.

This small device is attached to the trunk of a tree with a good view of the feeding station, at a height sufficient to protect it against theft and vandalism. An infrared motion sensor with adjustable sensitivity is used to detect movements above a certain threshold. The detected movements are evaluated by a microcontroller, which sends short message service (SMS) messages to a previously defined telephone number or e-mail address via a mobile telephone connected to the unit. Electrical power is provided by an adjustable solar panel and capacitors with a total capacity of 300 farads. This arrangement is sufficient to guarantee reliable, maintenance-free operation for many years, even at temperatures below freezing. This would not be possible with conventional rechargeable batteries. The system components are controlled by an upgraded version of the SMST4PIC module for transmitting and receiving SMS messages, which is a product of my design efforts.

After the unit had been operating successfully for more than 6 months, recurrent problems with the mobile telephone began to appear. The presumed cause of the problem was found after the unit was dismounted and analysed in the shop: the message 'SIM card error' appeared on the display of the telephone. The SIM was replaced by a new one, and the unit again worked perfectly for many months. However, the same pattern of errors started appearing again after around 10 months. As this behaviour was entirely at odds with the experience of mobile phone users, the problem was very puzzling. This was also why I came to learn about the situation.

When the SIM was fitted in a different mobile phone as a test, a similar error message appeared on the display. This appeared to be a rather clear indication that the SIM was responsible for the problem. The interesting question was to find out why the mobile phone was displaying this error message. A quick test of the SIM in a simple card terminal attached to a PC did not reveal any recognizable cause of the fault. The SIM transmitted a correct ATR after the power-up sequence, and it was possible to select the MF, the DFs, and some randomly chosen EFs. The conclusion was clearly that the card was basically in good working order.

The next step of the analysis was to monitor communications between the mobile phone and the SIM. This is quite easy to do with a few electronic components and a PC. There are also professional tools that provide detailed portrayals of the physical and logical layers of the data transmission process, but they are outside the range of my budget. In this case the jury-rigged arrangement was adequate, since it quickly became apparent that the error message occurred immediately after the VERIFY command was issued during PIN verification. The command sequence terminated with a '9240' return code instead of the usual '9000' return code, which meant there was some sort of memory problem.

A check of the command sequence showed that the microcontroller in the control module passed the PIN correctly to the mobile phone using the 'AT+CPIN' AT command. It was also coded as '1234' ('31 32 33 34 FF FF FF FF'), as is common among developers. PIN code verification was enabled to prevent the SIM from being used for other purposes if the unit was stolen. As the card was a normal subscriber card instead of a prepaid card, this was a quite understandable precaution, since there are periods in system operation during which stolen SIMs can go unnoticed for an extended length of time.

However, the fact that PIN verification was enabled could not have been the root cause of an EEPROM write error, since this is a common situation with manually operated mobile phones in many countries.

The cause lay in the software of the control unit microcontroller. In order to save energy, it switched off the mobile phone as soon as it was no longer needed. If a movement was detected, the microcontroller switched the telephone on again and then performed a PIN verification right away so it could log in to the network.

However, it waited until a certain number of movements had been detected before sending a message to the hunter, and this number was a configurable parameter. If the parameter value was set to 100, for instance, no message would be sent until 100 movements had been detected. This configuration capability was necessary to avoid overwhelming the hunter with a flood of SMS messages.

Unfortunately, the software mistakenly issued a PIN-verification AT command for every detected movement, even though this was not necessary. This caused the smart card to perform an enormous number of PIN verifications. Just imagine how many movements would be detected (with corresponding PIN verifications) if a group of wild boars visited the feeding station! It later turned out that this incorrect control of the mobile telephone was an overlooked relic of the time when the unit was still in the development stage.

To understand why this was a problem, you need to understand what happens during PIN verification in a smart card. Before comparing the PIN transferred by the command with the stored PIN, the smart card operating system increments the PIN error counter by 1, and then it performs the PIN comparison. Depending on the result of the comparison, it either resets the error counter to 0 or leaves it with the incremented value. These steps are necessary to prevent potential attacks based on interrupting power to the smart card.[1] However, they also result in two write accesses to the EEPROM page containing the error counter for each successful PIN verification.

There is a physical limit to the number of write accesses to the EEPROM. A typical specification is 500 000 guaranteed write accesses per EEPROM page. If PIN verifications are performed extremely often, it can happen that the error counter can no longer be written correctly after several hundred days of use. The SIM reports this situation to the mobile telephone by sending a '9240' return code, and the user receives a message that simply states that a card error occurred.

The actual cause of the premature failure of the SIMs had thus been found. The remedy consisted of making a small modification to the control program, after which the cards in the mobile telephones could reasonably be expected to have a service life of 10 years or more.

---

[1] See Rankl and Effing (2002)

The application described above is admittedly rather unusual, but the problem it illustrates arises again and again in automated data acquisition stations that use mobile telephones for data transmission and it is by no means limited to a few inexperienced hobbyists. If you are using a smart card either directly or indirectly, as in the example described here, is important to have a clear understanding of at least the basic aspects of the activities that occur in the operating system of a smart card. Otherwise you can easily find yourself confronted by otherwise avoidable difficulties in your applications.

Bibliography

Abelson H, Anderson R, Bellovin SM, Benaloh J, Blaze M, Diffei W, Gilmore J, Neumann PG, Rivest RL, Schiller JI, and Schneier B 1997 *The Risks of Key Recovery, Key Escrow, and Trusted Third-Party Encryption*. www.crypto.com.

Anderson RJ 2001 *Security Engineering*. John Wiley & Sons.

Anderson RJ and Needham RM 1995 Programming satan's computer, *Computer Science Today*. www.computersciencetoday.com.

Ausfuhrliste (Export Control List) 2004 *Ausfuhrliste: Anlage AL zur Außenwirtschaftsverordnung*, 25 May 2004. Bundesamt für Wirtschaft und Ausfuhrkontrolle.

AWG 2004 Außenwirtschaftsgesetz (Foreign Trade Act), 23 December 2004.

AWV 2001 Außenwirtschaftsverordnung (Foreign Trade Ordinance), 2 July 2001.

BAFA – Bundesamt für Wirtschaft und Ausfuhrkontrolle (Federal Office of Economics and Export Control). www.bafa.de.

BDSG 2001 Bundesdatenschutzgesetz (Federal Data Protection Act), 11 May 2001.

BfD – Bundesbeauftragte für den Datenschutz (Federal Commissioner for Data Protection). www.bfd.bund.de.

BNA – Bundesnetzagentur (Federal Network Agency). www.bundesnetzagentur.de.

Boehm BW 1981 *Software Engineering Economics*. Prentice Hall.

Bono S, Green M, Stubblefield A, Juels A, Rubin A, and Szydlo M 2005 *Security Analysis of a Cryptographically-Enabled RFID Device*. The Johns Hopkins University Information Security Institute, Baltimore.

BSI – Bundesamt für Sicherheit in der Informationstechnik (Federal Office for Information Security). www.bsi.de.

Chen Z 2000 *Java Card Technology for Smart Cards*. Addison Wesley.

Smart Card Applications: Design Models for using and programming smart cards   W. Rankl
© 2007 John Wiley & Sons, Ltd

CLUSIF 2002 *An Overview of Cyber-Crime in 2001.* Club de la Sécurité des Systémes d'Information Français, Paris.

Criteria 2005 Bekanntmachung zur elektronischen Signatur nach dem Signaturgesetz und der Signaturverordnung ('Notice Regarding Electronic Signatures Compliant with the Signature Act and the Signature Ordinance') (summary of suitable algorithms). Bundesnetzagentur (Federal Network Agency)

Crypto 2002 *Cryptography and Liberty: An International Survey of Encryption Policy.* Electronic Privacy Information Center, Washington, DC.

CWA 2004 *Application Interface for Smart Cards used as Secure Signature Creation Devices.* CWA 14890.

Datenschutz (Data Protection) 1996 *Anforderungen zur informationstechnischen Sicherheit bei Chipkarten* ('Requirements for Information Security with Regard to Chip Cards'). The Data Protection Commissioner of Hamburg, Hamburg.

ECR 2000 Council Regulation (EC) No 1334/2000 of 22 June 2000 setting up a Community regime for the control of exports of dual-use items and technology.

EMV Book 1 2004 *EMV Integrated Circuit Card Specification for Payment Systems, Book 1: Application Independent ICC to Terminal Interface Requirements,* Version 4.1. www.emvco.com.

EMV Book 2 2004 *EMV Integrated Circuit Card Specification for Payment Systems, Book 2: Security and Key Management,* Version 4.1. www.emvco.com.

EMV Book 3 2004 *EMV Integrated Circuit Card Specification for Payment Systems, Book 3: Application Specification,* Version 4.1. www.emvco.com.

EMV Book 4 2004 *EMV Integrated Circuit Card Specification for Payment Systems, Book 4: Cardholder, Attendant and Acquirer Interface Requirements,* Version 4.1. www.emvco.com.

EN 1546:2000 *Identification Card Systems: Inter-Sector Electronic Purse.*

ETSI – European Telecommunications Standards Institute. www.etsi.org.

EU 1995 *Directive 95/46/EC of the European Parliament and of the Council of 24 October 1995 on the Protection of Individuals with Regard to the Processing of Personal Data and on the Free Movement of Such Data. Official Journal of the European Communities,* L. 281, 23 November.

Finkenzeller F 1999 *RFID Handbook.* John Wiley & Sons.

Gamma E, Helm R, and Johnson RE 1994 *Design Patterns Elements of Reusable Object-Oriented Software.* Addison-Wesley.

Garstka H 2003 Informationelle Selbstbestimmung und Datenschutz, in Schulzki-Haddouti C *Bürgerrechte im Netz.* Bundeszentrale für politische Bildung, Bonn.

Global Platform 2003 *Open Platform: Card Specification*, Version 2.1.1.
www.globalplatform.org.

Haghiri Y and Tarantino T 2002 *Smart Card Manufacturing: A Practical Guide*. John
Wiley & Sons.

Hassler V, Manninger M, Gordeev M, and Muller M 2002 *Java Card for E-Payment
Applications*. Artech House, London.

Hunt A and Thomas D 1999 *The Pragmatic Programmer: From Journeyman to Master*.
Addison-Wesley.

Horster P and Fox D (ed.) 1999 *Datenschutz und Datensicherheit*. Vieweg Verlag,
Braunschweig.

ICAO 2002 *ICAO Machine Readable Travel Documents, Part 3: Size 1 and Size 2
Machine Readable Official Travel Documents*, 2nd edn, Doc 9303. www.icao.int.

ISO/IEC 7810:2003 *Identification Cards: Physical Characteristics*.

ISO/IEC 7811-1:2002 *Identification Cards: Recording Technique: Part 1 Embossing*.

ISO/IEC 7811-2:2001 *Identification Cards: Recording Technique: Part 2 Magnetic
Stripe: Low Coercivity*.

ISO/IEC 7813:2001 *Identification Cards: Financial Transaction Cards*.

ISO/IEC 7816-3:1997 *Identification Cards: Integrated Circuit(s) Cards: Part 3 Cards
with Contacts: Electrical Interface and Transmission Protocols*.

ISO/IEC 7816-4:2005 *Identification Cards: Integrated Circuit Cards: Part 4
Organization, Security and Commands for Interchange*.

ISO/IEC 7816-6:2004 *Identification Cards: Integrated Circuit Cards: Part 6
Interindustry Data Elements for Interchange*.

ISO/IEC 7816-8:2004 *Identification Cards: Integrated Circuit Cards: Part 8 Commands
for Security Operations*.

ISO/IEC 7816-9:2004 *Identification Cards: Integrated Circuit Cards: Part 9 Commands
for Card Management*.

ISO/IEC 7816-12:2005 *Identification Cards: Integrated Circuit Cards: Part 12 Cards
with Contacts: USB Electrical Interface and Operating Procedures*.

ISO/IEC 7816-15:2004 *Identification Cards: Integrated Circuit Cards: Part 15
Cryptographic Information Application*.

ISO 8402: 1994 *Quality Management and Quality Assurance: Vocabulary*.

ISO/IEC 8824: 2002 *Information Technology: Abstract Syntax Notation One (ASN.1)*.

ISO/IEC 8825: 2002 *Information Technology: ASN.1 Encoding Rules Specification of Basic Encoding Rules (BER), Canonical Encoding Rules (CER) and Distinguished Encoding Rules (DER)*.

ISO/IEC 14443-4:2001 *Identification Cards: Contactless Integrated Circuit(s) Cards: Proximity Cards*.

ITGH 2004 *IT-Grundschutzhandbuch*. Bundesanzeiger-Verlag, Cologne.

ITU X.509:2000 *Information Technology: Open Systems Interconnection: The Directory Authentication Framework*. www.itu.int.

JCAPN 2003 *Java Card Platform: Application Programming Notes*, Version 2.2.1, Sun Microsystems, Santa Clara, CA.

JCAPI 2003 *Java Card Platform: Application Programming Interface*, Version 2.2.1, Sun Microsystems, Santa Clara, CA.

JCRES 2003 *Java Card Platform: Runtime Environment Specification*, Version 2.2.1, Sun Microsystems, Santa Clara, CA.

JCVMS 2003 *Java Card Platform: Virtual Machine Specification*, Version 2.2.1, Sun Microsystems, Santa Clara, CA.

Lamport L 1981 Password authentication with insecure communication, *Communications of the ACM* 24, 11. ACM (Association for Computing Machinery), San Diego, CA.

Liggesmeyer P 2002 *Software-Qualität*. Spektrum Verlag, Heidelberg.

McConnell S 2002 *Code Complete*, 2nd edn. Barnes & Noble.

Menezes AJ, van Oorschot PC, and Vanstone SA 1997 *Handbook of Applied Cryptography*. CRC Press, Boca Raton, FL.

Open Card 2001 *Open Platform Card Specification*, Version 2.1, Open Card Foundation.

PCSC 2004 *PC/SC Interoperability Specification for ICCs and Personal Computer Systems*, V 2.00.11. www.smartcardsys.com.

PKCS #15 2000 *Cryptographic Token Information Format Standard*, V 1.1. www.rsa.com.

Rankl W and Effing W 2002 *Smart Card Handbook*, 3rd edn. John Wiley & Sons.

RFC 2289 1998 *A One-Time Password System*.

SATSA 2004 *Security and Trust Services API for Java 2 Platform Micro Edition Java Community Process (JCP)*, Version 1.0. jcp.org.

Schaar P 2002 *Datenschutz im Internet*. C.H. Beck, Munich.

Schneier B 1996 *Angewandte Kryptographie*. John Wiley & Sons.

Spillner A and Linz T 2003 *Basiswissen Softwaretest*. Dpunkt Verlag, Heidelberg.

TS 101 476:2002 *Digital cellular telecommunications system (Phase 2+): Subscriber Identity Module Application Programming Interface (SIM API): SIM API for Java CardTM:; Stage 2*, V8.5.0. ETSI.

TS 102 221:2005 *Smart cards: UICC–Terminal interface: Physical and logical characteristics*, Release 6, V6.8.0. ETSI.

TS 102 222:2005 *Integrated Circuit Cards (ICC): Administrative commands for telecommunications applications*, Release 6, V6.8.0. ETSI.

TS 31.102:2003 *3rd Generation Partnership Project: Technical Specification Group Terminals: Characteristics of the USIM application*, Release 6, V6.4.0. 3GPP.

TS 51.011:2003 *3rd Generation Partnership Project; Technical Specification Group Terminals; Specification of the Subscriber Identity Module – Mobile Equipment (SIM–ME) interface*, Release 4, V4.2.0. 3GPP.

TS 51.014:2003 *3rd Generation Partnership Project: Technical Specification Group Terminals: Specification of the SIM Application Toolkit for the Subscriber Identity Module – Mobile Equipment (SIM–ME) interface*, Release 4, V4.3.0. 3GPP.

V Model XT 2004 *V-Modell XT*. Federal Republic of Germany, www.v-modell-xt.de.

Wassenaar Arrangement: List of Dual Use Goods and Technologies And Munitions List. Vienna. 2004. www.wassenaar.org.

Wassenaar Arrangement on Export Controls for Conventional Arms and Dual-Use Goods and Technologies. www.wassenaar.org.

X.509:2000 *Information Technology: Open Systems Interconnection: The Directory Authentication Framework*. ITU.

Index